Hidden Markov Models in Finance

***Recent titles in the* INTERNATIONAL SERIES IN OPERATIONS RESEARCH & MANAGEMENT SCIENCE**

Frederick S. Hillier, Series Editor, *Stanford University*

Gass & Assad/ *AN ANNOTATED TIMELINE OF OPERATIONS RESEARCH: An Informal History*
Greenberg/ *TUTORIALS ON EMERGING METHODOLOGIES AND APPLICATIONS IN OPERATIONS RESEARCH*
Weber/ *UNCERTAINTY IN THE ELECTRIC POWER INDUSTRY: Methods and Models for Decision Support*
Figueira, Greco & Ehrgott/ *MULTIPLE CRITERIA DECISION ANALYSIS: State of the Art Surveys*
Reveliotis/ *REAL-TIME MANAGEMENT OF RESOURCE ALLOCATIONS SYSTEMS: A Discrete Event Systems Approach*
Kall & Mayer/ *STOCHASTIC LINEAR PROGRAMMING: Models, Theory, and Computation*
Sethi, Yan & Zhang/ *INVENTORY AND SUPPLY CHAIN MANAGEMENT WITH FORECAST UPDATES*
Cox/ *QUANTITATIVE HEALTH RISK ANALYSIS METHODS: Modeling the Human Health Impacts of Antibiotics Used in Food Animals*
Ching & Ng/ *MARKOV CHAINS: Models, Algorithms and Applications*
Li & Sun/ *NONLINEAR INTEGER PROGRAMMING*
Kaliszewski/ *SOFT COMPUTING FOR COMPLEX MULTIPLE CRITERIA DECISION MAKING*
Bouyssou et al./ *EVALUATION AND DECISION MODELS WITH MULTIPLE CRITERIA: Stepping stones for the analyst*
Blecker & Friedrich/ *MASS CUSTOMIZATION: Challenges and Solutions*
Appa, Pitsoulis & Williams/ *HANDBOOK ON MODELLING FOR DISCRETE OPTIMIZATION*
Herrmann/ *HANDBOOK OF PRODUCTION SCHEDULING*
Axsäter/ *INVENTORY CONTROL, 2nd Ed.*
Hall/ *PATIENT FLOW: Reducing Delay in Healthcare Delivery*
Józefowska & Węglarz/ *PERSPECTIVES IN MODERN PROJECT SCHEDULING*
Tian & Zhang/ *VACATION QUEUEING MODELS: Theory and Applications*
Yan, Yin & Zhang/ *STOCHASTIC PROCESSES, OPTIMIZATION, AND CONTROL THEORY APPLICATIONS IN FINANCIAL ENGINEERING, QUEUEING NETWORKS, AND MANUFACTURING SYSTEMS*
Saaty & Vargas/ *DECISION MAKING WITH THE ANALYTIC NETWORK PROCESS: Economic, Political, Social & Technological Applications w. Benefits, Opportunities, Costs & Risks*
Yu/ *TECHNOLOGY PORTFOLIO PLANNING AND MANAGEMENT: Practical Concepts and Tools*
Kandiller/ *PRINCIPLES OF MATHEMATICS IN OPERATIONS RESEARCH*
Lee & Lee/ *BUILDING SUPPLY CHAIN EXCELLENCE IN EMERGING ECONOMIES*
Weintraub/ *MANAGEMENT OF NATURAL RESOURCES: A Handbook of Operations Research Models, Algorithms, and Implementations*
Hooker/ *INTEGRATED METHODS FOR OPTIMIZATION*
Dawande et al./ *THROUGHPUT OPTIMIZATION IN ROBOTIC CELLS*
Friesz/ *NETWORK SCIENCE, NONLINEAR SCIENCE AND DYNAMIC GAME THEORY APPLIED TO THE STUDY OF INFRASTRUCTURE SYSTEMS*
Cai, Sha & Wong/ *TIME-VARYING NETWORK OPTIMIZATION*

***** *A list of the early publications in the series is at the end of the book* *****

Hidden Markov Models in Finance

Edited by

Rogemar S. Mamon
Robert J. Elliott

Springer

Rogemar S. Mamon
University of Western Ontario
London, Canada

Robert J. Elliott
University of Calgary
Calgary, Canada

ISBN-13: 978-3-4419-4380-4
ISBN-13: 978-0-387-71163-8 (e-book)

Printed on acid-free paper.

9 8 7 6 5 4 3 2 1

springer.com

Contents

List of Contributors

Abdul Abiad
International Monetary Fund
700 19th St. NW, Washington,
DC 20431 USA
aabiad@imf.org

Lakhdar Aggoun
Department of Mathematics and Statistics
Sultan Qaboos University
P.O.Box 36, Al-Khod 123,
Sultanate of Oman
laggoun@squ.edu.om

Robert J. Elliott
Haskayne School of Business
University of Calgary
2500 University Drive NW,
Calgary, Alberta, Canada T2N 1N4
relliott@ucalgary.ca

Malgorzata W. Korolkiewicz
School of Mathematics and Statistics
University of South Australia
Adelaide,
South Australia 5095
malgorzata.korolkiewicz@unisa.edu.au

Shangzhen Luo
Department of Mathematics,
University of Northern Iowa
Cedar Falls, IA 50614-0506
USA
luos@uni.edu

Rogemar S. Mamon
Department of Statistical and Actuarial Sciences
The University of Western Ontario
London, Ontario,
Canada N6A 5B7
rmamon@stats.uwo.ca

Emilio Russo
Department of Mathematics, Statistics, Computing & Applications and Faculty of Economics and Business Administration
University of Bergamo
via Salvecchio 19,
24129 Bergamo, Italy
emilio.russo@unibg.it

Kin Bong Siu
Department of Statistics and Actuarial Science
The University of Hong Kong
Pokfulam Road, Hong Kong
h0010297@hkusua.hku.hk

Tak Kuen Siu
Department of Actuarial Mathematics and Statistics
School of Mathematical and Computer Sciences,
Heriot-Watt University, Edinburgh, UK EH14 4AS
T.K.Siu@ma.hw.ac.uk

Fabio Spagnolo
CARISMA, Business School
Brunel University, Uxbridge, UK UB8 3PH
Fabio.Spagnolo@brunel.ac.uk

Anatoly V. Swishchuk
Department of Mathematics and Statistics
University of Calgary
2500 University Drive NW,
Calgary, Alberta, Canada T2N 1N4
aswish@math.ucalgary.ca

Allanus H. Tsoi
Department of Mathematics
University of Missouri
Columbia, Missouri 65211
USA
tsoi@math.missouri.edu

Craig A. Wilson
College of Commerce
University of Saskatchewan
Saskatoon, Saskatchewan,
Canada S7N 5A7
cwilson@commerce.usask.ca

Shu Wu
Department of Economics
the University of Kansas
Lawrence, KS 66045, USA
shuwu@ku.edu

Hailiang Yang
Department of Statistics and Actuarial Science
The University of Hong Kong
Pokfulam Road, Hong Kong
hlyang@hkusua.hku.hk

Yong Zeng
Department of Mathematics and Statistics
University of Missouri at Kansas City
Kansas City, MO 64110, USA
zeng@mendota.umkc.edu

Biographical Notes

Editors

Rogemar S. Mamon is presently a faculty member of the Department of Statistical and Actuarial Sciences at the University of Western Ontario (UWO), London, Ontario, Canada. Prior to joining UWO he was Research Lecturer B in Financial Engineering at Brunel University, West London, England; Assistant Professor of Statistics at the University of British Columbia; and Assistant Professor of Statistics and Actuarial Science at the University of Waterloo. His publications have appeared in various peer-reviewed journals in statistics, applied mathematics, quantitative finance and mathematical education. He is a Registered Practitioner of the Higher Education Academy UK, a Chartered Mathematician of the Institute of Mathematics and its Applications, and a Chartered Scientist of the UK Science Council.

Robert Elliott is the RBC Financial Group Professor of Finance at the University of Calgary, Canada. He has authored and co-authored more than 300 refereed articles covering the subject of harmonic and functional analysis, hypoelliptic operator, differential games, stochastic calculus and control, stochastic filtering theory and more recently, quantitative finance. His well-known books include Stochastic Calculus (1982), a Springer-Verlag graduate text; Hidden Markov Models: Estimation and Control (1994) published by Springer-Verlag, co-authored with L. Aggoun and B. Moore; Measure Theory and Filtering (2004), published by Cambridge University Press, co-authored with L. Aggoun; Mathematics of Financial Markets, 1st ed., 1999, 2nd ed., 2005 published by Springer-Verlag, co-authored with P. Kopp; and Binomial Methods in Finance (2005) published by Springer-Verlag, co-authored with John van der Hoek. He held previous academic positions at Newcastle-upon-Tyne, Oxford, Warwick and Hull in the UK; Yale, Northwestern, and Brown in the USA; and Adelaide in Australia. He is a Professor Emeritus in Mathematics and held the AF Collins Professorial Chair in Finance from 1999 to

2002 at the University of Alberta. He received first class BA and MA degrees from Oxford and PhD and DSc degrees from Cambridge.

Contributors

Abdul Abiad is an Economist in the European Department of the International Monetary Fund (IMF). He was previously with the IMF's Research Department, where he worked on early warning systems for currency crises, as well as on issues relating to the financial sector. He has a PhD from the University of Pennsylvania, where his dissertation focused on the application of Markov models to early warning systems for currency crises.

Lakhdar Aggoun is an Associate Professor of Applied Probability at Sultan Qaboos University, Oman. He has published several articles in filtering and control theory. He is a co-author of two books with Robert Elliott published by Springer-Verlag and Cambridge University Press. He holds a B.Sc. in Mathematics from the University of Constantine, Algeria; an MSc in Mathematics from Stevens Institute of Technology, USA; and an MSc in Probability and a PhD in Applied Probability both from the University of Alberta, Canada.

Malgorzata Korolkiewicz is currently a Lecturer at the School of Mathematics and Statistics and a Researcher at the Centre for Industrial and Applied Mathematics at the University of South Australia. Her research interests include mathematical finance, derivatives markets and credit risk. She received her PhD from the University of Alberta, Canada.

Shangzhen Luo is an Assistant Professor in the Department of Mathematics at the University of Northern Iowa (UNI). He joined the UNI faculty after completing his Ph.D. in mathematics from the University of Missouri-Columbia. He received his M.Phil. degree in mathematics from Hong Kong University of Science and Technology and B.S. in mathematics from Nankai University. His research interests include hidden Markov models, stochastic control, classical risk theory and their applications to finance and insurance.

Emilio Russo is currently a PhD student at the University of Bergamo, Italy. His main area of interests are option pricing, optimal asset allocation in continuous and discrete time, insurance policies and modelling interest and exchange rates. He holds an MPhil in mathematical finance from Brunel University, UK and bachelour's degree in statistics and actuarial sciences from the University of Calabria, Italy. He has some publications in the field of option pricing and asset allocation. Recently, he is working on valuation models for equity-linked insurance policies as part of his doctoral dissertation on the subject of computational methods in economics, financial decisions and forecasting.

Bonny K.B. Siu is a part-time graduate student at the Department of Statistics and Actuarial Science of the University of Hong Kong, and is a Customer Analytics Assistant Manager in Hong Kong and Shanghai Banking Corporation (HSBC). During the last few years he has been involved in many marketing sales campaigns and customer relationship management system development in the Asia-Pacific region and is currently located at the company's Hong Kong head office. His main area of interest is credit risk modelling.

Tak Kuen Siu is a Lecturer in actuarial mathematics in the Department of Actuarial Mathematics and Statistics at Heriot-Watt University, Scotland, where he teaches courses in life insurance mathematics and financial mathematics. His research interests are mathematical finance and actuarial science with specialisation in risk measurement and management, actuarial methods for pricing derivatives and insurance products, credit risk models and financial time series analysis. He has published research papers in international refereed academic and professional journals in actuarial science and mathematical finance, including *Insurance: Mathematics and Economics, North American Actuarial Journal, Quantitative Finance, Risk Magazine, International Journal of Theoretical and Applied Finance and Applied Mathematical Finance.* He received his BSc in mathematics from Hong Kong University of Science and Technology in 1998 and PhD in statistics and actuarial science from the University of Hong Kong in 2001.

Fabio Spagnolo is a Reader in Finance at Brunel Business School, UK. He has published over a dozen of articles in the area of econometrics and finance in academic journals including *Economics Letters, Journal of Applied Econometrics, Journal of Econometrics, Journal of Forecasting* and *Studies in Nonlinear Dynamics & Econometrics.*

Anatoliy Swishchuk is an Assistant Professor of Mathematical Finance at the University of Calgary, Canada. His research interests include the modelling and pricing of various swaps, option pricing, regime-switching models and stochastic models with delay in finance. He is the author of many research papers and 8 books. He received his PhD and DSc from the Institute of Mathematics, Kiev, Ukraine.

Allanus Tsoi obtained his BSc at the University of Washington, his M.Sc. at the University of Illinois at Urbana-Champaign, and his Ph.D. at the University of Alberta, Canada. His research area in mathematical finance includes American options and Asian options, stochastic volatility dynamics and estimation, financial applications of hidden Markov filtering techniques, modelling term structure of stochastic interest rates and analysis of their derivatives, as well as financial applications of white noise theory. He is at present an Associate Professor of Mathematics at the University of Missouri at Columbia, USA. He played an essential role in setting up the mathematical finance program at the University of Missouri-Columbia.

Craig Wilson is an Associate Professor of Finance at the University of Saskatchewan, Canada. He researches in the areas of fixed income securities, asset pricing, and financial investments. In particular, he combines methods from the theory of stochastic processes and econometrics to model and analyse financial situations. He received his BSc in mathematics, BCom in finance, and PhD in finance degrees from the University of Alberta, Canada.

Shu Wu is an Assistant Professor and Oswald Scholar in the Department of Economics at the University of Kansas. His main research areas are macroeconomics and finance. He has published in the *Journal of Monetary Economics, Journal of Money, Credit and Banking, Macroeconomic Dynamics, Annals of Finance* and *International Journal of Theoretical and Applied Finance,* amongst others. He received his PhD from Stanford University in 2000.

Hailiang Yang is an Associate Professor in the Department of Statistics and Actuarial Science of the University of Hong Kong. His current research interests are actuarial science and mathematical finance. He has published over 80 papers on actuarial science, mathematical finance and stochastic control in academic journals including *Mathematical Finance, ASTIN Bulletin, Insurance: Mathematics and Economics, Scandinavian Actuarial Journal, Stochastic Processes and their Applications, Journal of Applied Probability, Advances of Applied Probability and IEEE Transactions on Automatic Control.* He is an associate editor of *Insurance: Mathematics and Economics.* He received his PhD from the University of Alberta, Canada.

Yong Zeng is an Associate Professor in the Department of Mathematics and Statistics at the University of Missouri, Kansas City. He is a visiting associate professor at the University of Tennessee at Knoxville (Fall 2006) and Princeton University (Spring 2007). He was also an invited visiting scholar in Wayne State University, University of Wisconsin - Madison and University of Alberta. His research interests include marked point processes; Bayesian inference (estimation, hypothesis testing and model selection) via filtering of ultra-high frequency data, filtering with marked point process observations, and particle filtering; modelling and estimating the term structure of interest rates with regime-switching risks; MLE and MCMC for jump diffusion process with discrete observations; and statistical modelling for network traffic data. He received his BS degree from Fudan University (Shanghai, China) in 1990, MS from the University of Georgia at Athens in 1994, and Ph.D. from University of Wisconsin at Madison in 1999; all degrees are in statistics.

Preface

The term *hidden Markov model* (HMM) is more familiar in the speech signal processing community and communication systems but recently it is gaining acceptance in finance, economics and management science. The term HMM is frequently restricted to models with states and measurements in a discrete set and in discrete time. However, there is no reason why these restrictions cannot be relaxed, and so one can extend the modelling in continuous time and include observations with continuous range. The theory of HMM deals with *estimation,* which involves signal filtering, model parameter identification, state estimation, signal smoothing, and signal prediction; and *control,* which refers to selecting actions which effect the signal-generating system in such a way as to achieve certain control objectives. In the HMM implementation, reference probability methods are employed. This is a set of procedures designed in the reformulation of the original estimation and control task in a fictitious world so that well-known results for identically and independent distributed random variables can be applied. Then the results are reinterpreted back to the real world with the original probability measure.

To get a better understanding of an HMM, consider a message sequence $X_k (k = 1, 2, \ldots)$ consisting of 0's and 1's depicted in Figure 1. Then, possibly the binary signal X (a Markov chain) is transmitted on a noisy communications channel such as a radio channel and the additive noise is illustrated in Figure 2. When the signal is detected at the receiver, we obtain some resultant Y_k. What we get therefore after combining Figures 1 and 2 is a binary Markov chain **hidden** in noise given in Figure 3.

Essentially, the goal is to develop optimal estimation algorithms for HMMs to filter out the random noise in the best possible way. HMM filtering theory, therefore, discusses the optimal recursive estimation of a noisy signal given a sequence of observations. In electrical engineering for example, one is interested to determine the charge $\mathcal{Q}(t)$, at time t at a fixed point in an electric circuit. However, due to error in the measurement of $\mathcal{Q}(s), (s < t)$ one cannot really measure $\mathcal{Q}(s)$, but rather just a noisy version of it. The objective is to

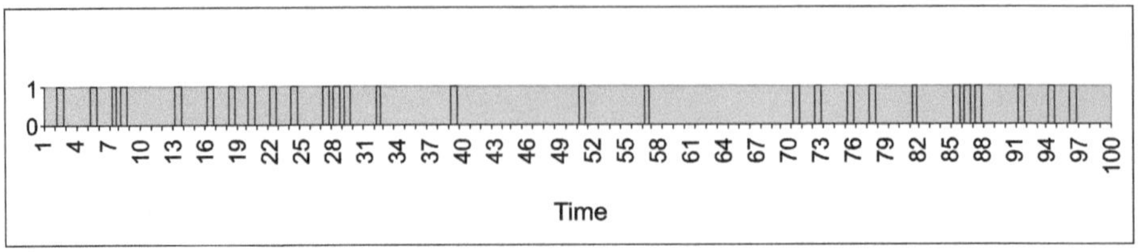

Fig. 1. Markov Chain

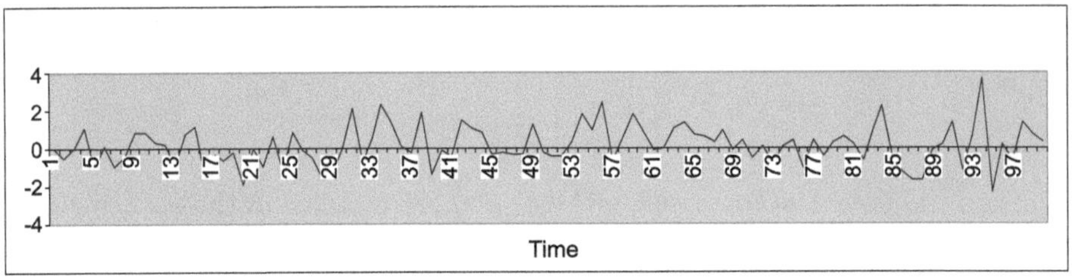

Fig. 2. Noise

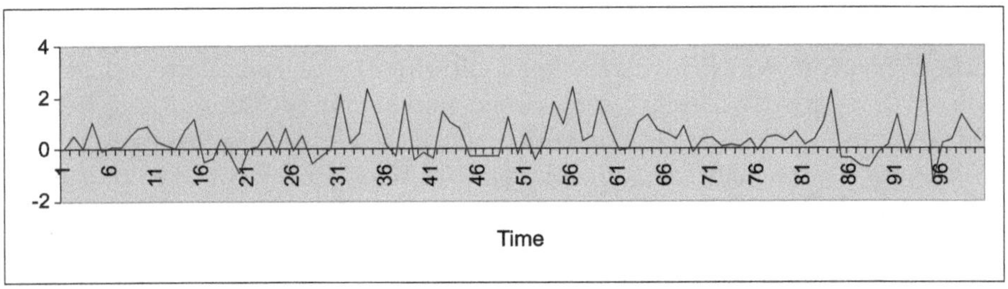

Fig. 3. Noisy Observations $Y(k)$

"filter" the noise out of our observations. In a similar manner, we might ask whether financial data, interest rates, asset price processes, exchange rates, commodity prices, etc. contain information about latent variables. If so, how might their behaviour in general and in particular their dynamics be estimated?

The use of HMMs is also motivated by significant empirical evidence from the literature that favours and endorses Markov-switching models in the study of many macroeconomic variables. This provides more flexibility to financial models and incorporates stochastic volatility in a simple way. Earlier development in this area during the late 80's within the time series context was pioneered by James Hamilton, amongst others, in a work that proposed to have the unobserved regime follow a Markov process. Indeed, examples of many models in which the shift of regimes is governed by a discrete or continuous time Markov chain abound in finance and economics in the areas of business cycles, stock prices, foreign exchange, interest rates and option valuation. The rationale behind the regime-switching framework is that the market may switch from time to time between, say, a "quiet" (stable low volatility) state and a "turbulent" (unstable high volatility) state. In general the Markov chain states can refer to any number of conceivable "state of the economy".

Within the HMM set-up and related modelling structures, this monograph offers a collection of papers dealing with the theory and empirical investigations probing the particular aspects of dynamic financial and economic modelling outlined above. The main themes in this collection include pricing, risk management, model calibration and parameter estimation.

This volume opens with two papers devoted to term structure of interest rates. In 'An exact solution of the term structure of interest rate under regime-switching risk', Shu Wu and Yong Zeng derive a closed-form solution to the term structure under an essentially affine-type model using log-linear approximation. It is shown that the market price of regime-switching risk affects the long-end of the yield curve and hence this is a significant component of the term premium for long-term bonds. Then Robert Elliott and Craig Wilson in 'The term structure of interest rates in a hidden Markov setting', develop an interest rate model whereby the stochastic nature of volatility and mean reversion is introduced in a simple and tractable way. Zero-coupon bond price is calculated. Empirical work using non-linear regression model illustrates that a 3-state Markov chain is able to explain considerably the dynamics of the yield rate data.

The theme of HMM regime-switching-based models continue with Tak Kuen Siu's paper that demonstrates the interplay of methodologies in finance and actuarial science to successfully price insurance products with recent innovations. In 'On Fair valuation of participating life insurance policies with regime switching', he employs a regime-switching Esscher transform to value insurance policies with embedded exotic features. The valuation is performed within the basic geometric Brownian motion model but whose drift and volatility parameters are modulated by a hidden Markov model. Under the same market framework, Robert Elliott and Anatoliy Swishchuk investigate the valuation of options and variance swaps.

Two contributions then tackle the measurement and management of financial risks. In 'Smoothed parameter estimation for a hidden Markov model of credit quality', Malgorzata Korolkiewicz and Robert Elliott propose a model for the evolution of companies' credit rating using a hidden Markov chain in discrete time. Smooth estimates for the state of the Markov chain and auxiliary parameters are also obtained. Kim Bong Siu and Hailang Yang, in 'Expected shortfall under a model with market and credit risks', present an integrated model that handle both credit and market risks. Two approaches in calculating VaR and ES are given: recursive equations and Monte Carlo methods. A weak Markov chain model is also outlined in their attempt to take into account the dependency of risks.

This is followed by papers on the filtering of HMM via change of probability measures. The development of general filters for the state, occupation time and total number of jumps of a weak Markov chain is examined by Shangzhen Luo and Allanus Tsoi in their article 'Filtering of hidden weak Markov chain-discrete range observations'. Weak Markov chains may be suitable in modelling financial and economic processes that exhibit some form of memory. The study of future demands and inventory level via a discrete HMM in discrete time is the focus of Lakhdar Aggoun's paper 'Filtering of a partially observed inventory system'. The recursive estimation of the joint distribution of the level of stock and actual demand together with the re-estimation of model parameters is highlighted.

The monograph culminates with two papers that explore a thought-provoking hypothesis and challenging questions in economics. Emilio Russo, Fabio Spagnolo and Rogemar S Mamon, in 'An empirical investigation of the unbiased forward exchange rate hypothesis in a regime-switching market', use a Markov chain to describe structural change brought about by the intervention of central banks and other changes in the monetary policies as well as test the validity of the unbiased forward exchange rate hypothesis using US dollar/UK sterling pound exchange rate data. Abdul Abiad put forward the use of Markov regime-switching model to identify and characterise currency crisis periods. In 'Early warning systems (EWS) for currency crises: A regime-switching approach', he provides empirical support that a regime-switching model outperforms standard EWS in signaling crises and reducing false alarms. Country-by-country analyses of data for the period 1972-1999 from five Asian countries (Indonesia, Korea, Malaysia, the Philippines and Thailand), all of which experienced currency crises, were conducted.

We hope that this monograph will provide more insights to the financial research community and open avenues for more interesting problems. Specifically, it is our hope that this volume will raise more stimulating questions for further discussions in our concerted effort to build dynamic models that could incorporate the important stylised features of a financial market and capture better the significant factors of an economy.

Rogemar S. Mamon (University of Western Ontario)
Robert J. Elliott (University of Calgary)

Acknowledgement. Rogemar Mamon acknowledges gratefully the financial support received from the British Academy (Grant No. OCG-41559), Brunel University (BRIEF Award No. 735) and the Faculty of Science, University of Western Ontario. Robert Elliott wishes to thank the Social Sciences and Humanities Research Council of Canada for the support of his research. The help and expertise of PhD candidate Luka Jalen (Brunel University, UK) in producing consistent latex files have been remarkable and invaluable. Special thanks go to Professor Gautam Mitra for initiating the idea of working on the main subject of this monograph and for his support in making this project materialised. Finally, both editors would like to express their sincere appreciation for the superb assistance of Fred Hillier, Gary Folven, Tracey Howard and Carolyn Ford, all at Springer, in the production of this monograph.

1

An Exact Solution of the Term Structure of Interest Rate Under Regime-Switching Risk

Shu Wu[1] and Yong Zeng[2]

[1] Department of Economics
the University of Kansas
Lawrence, KS 66045, USA
shuwu@ku.edu

[2] Department of Mathematics and Statistics
University of Missouri at Kansas City
Kansas City, MO 64110, USA
zeng@mendota.umkc.edu

Summary. Regime-switching risk has been recently studied in an general equilibrium setting and empirically documented as an significant factor in bond premium. In this paper we apply no arbitrage approach to derive an exact solution of the term structure of interest rates in an essentially-affine-type model under regime-switching risk.

Key words: Term structure model, regime-switching risk, marked point process, affine diffusion

1.1 Introduction

Much documented empirical evidence implies that the aggregate economy has recurrent shifts between distinct regimes of the business cycle (e.g Hamilton [17], and Diebold and Rudebusch [10]). These results have motivated the recent studies of the impact of regime shifts on the entire yield curve using dynamic term structure models. A common approach is to incorporate Markov-switching (or hidden Markov chains) into the stochastic processes of the pricing kernel and/or state variables. Indeed the regime-dependence offers greater econometric flexibilities in empirical models of the term structure such as Bansal and Zhou [1]. However, as pointed out by Dai and Singleton [8], the risk of regime shifts is not priced in many of these models, and hence it does not contribute independently to bond risk premiums.

Without pricing the risk of regime shifts, the previous studies have essentially treated the regime shifts as an idiosyncratic risk that can be diversified away by bond investors. However, Bansal and Zhou [1] and Wu and Zeng [22] empirically showed that regimes are intimately related to the business cycle, suggesting a close link between the regime shift and aggregate uncertainties.

Extending the aforementioned strand of literature, Wu and Zeng [22] develop a dynamic term structure model under the systematic risk of regime shifts in a general equilibrium setting similar to Cox, Ingersoll and Ross [5] [6] (henceforth CIR). The model implies that bond risk premiums include two components under regime shifts: (i) a regime-dependent risk premium due to diffusion risk as in the previous studies, and (ii) a regime-switching risk premium that depends on the covariations between the discrete changes in marginal utility and bond prices across different regimes. This new component of the term premiums is associated with the systematic risk of recurrent shifts in bond prices (or interest rates) due to regime changes and is an important factor that affects bond returns. Furthermore, we also obtain a closed-form solution of the term structure of interest rates under an affine-type model using the log-linear approximation similar to that in Bansal and Zhou [1]. The model is estimated using the Efficient Method of Moments applied to monthly data on 6-month treasury bills and 5-year treasury bonds from 1964 to 2000. We find that the market price of regime-switching risk is highly significant and affects mostly the long-end of the yield curve. The regime-switching risk, as expected, accounts for a significant portion of the term premiums for long-term bond.

A drawback in Wu and Zeng [22] is that in an affine-type model, the closed-form solution of the yield curve is obtained under log-linear approximation[3]. In this paper, using no-arbitrage approach, we derive an exact solution of the term structure of interest rates in a more general essentially-affine-type model under regime-switching risk.

In the standard affine models (such as Duffie and Kan [13] and Dai and Singleton [7]), the market price of diffusion risk is proportional to the volatility of the state variable. Such a structure guarantees that the models satisfy a requirement of no-arbitrage; risk compensation goes to zero as risk goes to zero. However, as Duffee [12] points out, this structure limits the variation of the compensations that investors anticipate to obtain when encountering a risk. More precisely, since the compensation is bounded below by zero, it cannot change sign over time. Duffee [12] argues that this is the main reason why the completely affine models fails at forecasting. He suggests a broarder class of essentially affine models to break the tight link between risk compensation and interest rate volatility. These more general models are shown to have better forecasting ability than the standard affine models. In this paper, we

[3] Due to the nature of log-linear approximation, we conjecture that the error bound should be in the order of r^2. Wu and Zeng [23] derived the closed-form for the multi-factor affine models with both jump and regime-switching risks using log-linear approximation.

introduce regime shifts into the class of essentially-affine models. Our model with exact solution presented here should prove useful in forecasting future yields.

To the best of our knowledge, three other papers also presented exact solutions for regime switching term structure models. Two of them are continuous-time models: one is Landen [19] and the other is Dai and Singleton [8]. Landen [19] focused on the case under risk-neutral probability measure, she did not mention anything about the market price of regime switching risk. Dai and Singleton [8] surveyed the theoretical specification of dynamic term structure models. Moreover, they proposed a Gaussian affine-type model with regime-switching risk and constant volatility within each regime. In our model, we allow for stochastic volatilities in each regime and the diffusion risk is in an essentially affine form. The third one is Dai, Singleton and Yang [9]. They develop and empirically implement a discrete-time Gaussian dynamic term structure model with priced factor and regime-shift risks.

The rest of the paper is organized as follows. Section 2 presents a simpler expressive form of regime-shifting using marked point process (or random measure) approach. Section 3 develops a framework for the term structure of interest rates with regime-switching risk using the no arbitrage approach. Section 4 specifies an essentially-affine-type model with regime switching risk and derives an exact solution. Section 5 concludes with some future research topics.

1.2 A new representation for modeling regime shift

In the literature of interest rate term structure, there are three approaches to model regime shifting process. The first approach is the *Hidden Markov Model*, summarized in the book of Elliott et al. [15], and its application to the term structure can be found in Elliott and Mamon [16]. The second approach is the *Conditional Markov Chain*, discussed in Yin and Zhang [24], and its applications to the term structure are in Bielecki and Rutkowski ([2],[3]). The third approach is the *Marked Point Process* or the *Random Measure* approach as in Landen [19]. Due to its notational simplicity, here, we follow the third approach but propose a new and simpler representation. In Landen [19], the mark space is a product space of regime $E = \{(i,j) : i \in \{1,...,N\}, j \in \{1,2,...,N\}, i \neq j\}$, including all possible regime switchings. Below, we simplify the mark space to the space of regime only and consequently simplify the corresponding random measure as well as the equation for $s(t)$ or s_t, which is defined as the most recent regime.

There are two steps to obtain the simple expression for $s(t)$.

Step 1: We define a random counting measure. Let the mark space $U = \{1,2,...,N\}$ be all possible regimes with the power σ-algebra, and u be a generic point in U. Let A be a subset of U. Let $m(t, A)$ counts the cumulative number of entering a regime that belongs to A during the time interval $(0, t]$. For example, $m(t, \{u\})$ counts the cumulative number of entering regime u during $(0, t]$. Note that m is a random counting measure. Let η be the usual counting measure on U. Then, η has the following two properties: for $A \in U$, $\eta(A) = \int I_A \eta(u)$ (i.e. $\eta(A)$ counts the number of elements in A) and $\int_A f(u)\eta(u) = \sum_{u \in A} f(u)$.

A marked point process or a random measure is uniquely characterized by its stochastic intensity kernel[4]. Let $x(t)$ denote a state variable to be defined later. Then, the stochastic intensity kernel of $m(t, \cdot)$ can be defined as

$$\gamma_m(dt, du) = h\big(u; x(t-), s(t-)\big)\eta(du)dt, \tag{1.1}$$

where $h(u; x(t-), s(t-))$ is the conditional regime-shift (from regime $s(t-)$ to u) intensity at time t (we assume $h(u; x(t-), s(t-))$ is bounded). Heuristically, $\gamma_m(dt, du)dt$ can be thought of as the conditional probability of shifting from regime $s(t-)$ to regime u during $[t, t + dt)$ given $x(t-)$ and $s(t-)$. Then, $\gamma_m(t, A)$, the compensator of $m(t, A)$, can be written as

$$\begin{aligned}\gamma_m(t, A) &= \int_0^t \int_A h\big(u; x(\tau-), s(\tau-)\big)\eta(du)d\tau \\ &= \sum_{u \in A} \int_0^t h\big(u; x(\tau-), s(\tau-)\big)d\tau.\end{aligned}$$

Step 2: We are in the position to present the integral and differential forms for the evolution of regime, $s(t)$, using the random measure defined above. First, the integral form is

$$s(t) = s(0) + \int_{[0,t]\times U} \big(u - s(\tau-)\big)m(d\tau, du). \tag{1.2}$$

Note that $m(d\tau, du)$ is zero most of the time and only becomes one at regime-switching time t_i with $u = s(t_i)$, the new regime at time t_i. Observe that the above expression is but a telescoping sum: $s(t) = s(0) + \sum_{t_i < t}(s(t_i) - s(t_{i-1})$. Second, the differential form is

$$ds(t) = \int_U \big(u - s(t-)\big)m(dt, du). \tag{1.3}$$

To see the above differential equation is valid, assuming that there is a regime-switching from $s(t-)$ to u which occurs at time t, then $s(t) - s(t-) = (u - s(t-))$ implying $s(t) = u$.

These two forms are crucial in the following two sections.

---

[4] See Last and Brandt [20] for detailed discussion of marked point process, stochastic intensity kernel and related results.

1.3 The model

1.3.1 Two state variables

We assume that there are two state variables. One describes the regime change, $s(t)$ or s_t, which stands for the most recent regime. As described in Section 2, s_t follows (1.2) or (1.3). The other state variables, x_t, is described by a diffusion

$$dx_t = \mu(x_t, s_t)dt + \sigma(x_t, s_t)dW_t \tag{1.4}$$

where the drift term and the diffusion term are in general time-varying and regime-dependent, and W_t is a standard Brownian motion.

The instantaneous short-term interest r_t is a linear function of x_t given s_t, i.e.,

$$r_t = \psi_0(s_t) + \psi_1 x_t \tag{1.5}$$

where $\psi_0(s_t)$ is a constant depending on regime but ψ_1 is not. When ψ_1 is also regime-dependent, we cannot obtain an exact solution.

1.3.2 Pricing kernel

Under certain technical conditions, the absence of arbitrage is sufficient for the existence of the pricing kernel (see Harrison and Kreps [18]). We further specify the pricing kernel M_t as

$$\begin{aligned}\frac{dM_t}{M_{t-}} &= -\, r_{t-}dt - \lambda_D(x_t, s_t)dW_t \\ &\quad - \int_U \lambda_S(u; x_t, s_{t-})\big[m(dt, du) - \gamma_m(dt, du)\big]\end{aligned} \tag{1.6}$$

where $\lambda_D(x_t, s_t)$ is the market price of diffusion risk, which is also regime-dependent; and $\lambda_S(u; x_{t-}, s_{t-})$ is the market price of regime-switching (from regime s_{t-} to regime u) risk given x_t and s_{t-}.

Note that the explicit solution for M_t can be obtained by Doleans-Dade exponential formula (Protter [21]) as the following:

$$\begin{aligned}M_t = &\left(e^{-\int_0^t r_\tau d\tau}\right) \left(e^{-\int_0^t \lambda_D(x_\tau, s_\tau)dW(\tau) - \frac{1}{2}\int_0^t \lambda_D^2(x_\tau, s_\tau)d\tau}\right) \times \\ &\left(e^{\int_0^t \int_U \lambda_S(u; s_{\tau-}, s_{\tau-})\gamma_m(d\tau, du) + \int_0^t \int_U \log(1-\lambda_S(u; x_{\tau-}, s_{\tau-}))m(d\tau, du)}\right).\end{aligned} \tag{1.7}$$

1.3.3 The risk-neutral probability measure

The specifications above complete the model for the term structure of interest rates, which can be solved by a change to the risk-neutral probability measure. We first obtain the following two lemmas. The first lemma characterizes the equivalent martingale measure under which the interest rate term structure is determined. The second lemma specifies the dynamics of the short rate and the regime under the equivalent martingale measure.

Lemma 1. *For fixed $T > 0$, the equivalent martingale measure* $\mathbf{Q}$ *is defined by the Radon-Nikodym derivative*

$$\frac{dQ}{dP} = \frac{\xi_T}{\xi_0}$$

where for $t \in [0,T]$

$$\begin{aligned}\xi_t = &\left(e^{-\int_0^t \lambda_D(x_\tau,s_\tau)dW(\tau)-\frac{1}{2}\int_0^t \lambda_D^2(x_\tau,s_\tau)d\tau}\right) \times \\ &\left(e^{\int_0^t \int_U \lambda_S(u;x_\tau,s_{\tau-})\gamma_m(d\tau,du)+\int_0^t \int_U \log(1-\lambda_S(u;x_\tau,s_{\tau-}))m(d\tau,du)}\right)\end{aligned} \tag{1.8}$$

provided λ_D, λ_S *and* h *in* $m(t,A)$ *are all bounded in* $[0,T]$.

Proof. Obviously, $\xi_t > 0$ for all $0 \le t \le T$. By Doleans-Dade exponential formula, ξ_t can be written in stochastic differential equation form as

$$\frac{d\xi_t}{\xi_t} = -\lambda_D(x_t,s_t)dW_t - \int_U \lambda_S(u;x_t,s_{t-})\big[m(dt,du) - \gamma_m(dt,du)\big]. \tag{1.9}$$

Since W_t and $m(t,A) - \gamma_m(t,A)$ are martingales under P, ξ_t is a local martingale.

Since ξ_t is a $\mathbf{P}$-local martingale and $\xi_0 = 1$, it suffices to show that $E([\xi]_t) < \infty$ to obtain $E(\xi_t) = 1$ for all t, because ξ becomes a martingale if $E([\xi]_t) < \infty$ for all t, where $[\xi]_t$ is the quadratic variation process of ξ. Let

$$K_t = -\int_0^t \lambda_D(x_\tau,s_\tau)dW_\tau - \int_0^t \int_U \lambda_S(u;x_\tau,s_{\tau-})[m(dt,du) - \gamma_m(d\tau,du)]$$

By assumption, we suppose that $|\lambda_D(x_t,s_t)| \le C_D$ for all x_t and s_t, and $|\lambda_S(u;x_t,s_{t-})| \le C_S$ and $h(u;x_t,s_{t-}) \le C_h$ for all u, x_t and s_{t-}. Using the properties of quadratic variation for semimartingales (see Section 2.6 of Protter [21]), we have

$$[K]_t = \int_0^t \lambda_D^2(x_\tau,s_\tau)d\tau + \int_0^t \int_U \lambda_S^2(u;x_\tau,s_{\tau-})m(d\tau,du).$$

Observe that for $t \le T$,

$$\int_0^t \lambda_D^2(x_\tau,s_\tau)d\tau \le C_D^2\ T$$

and

$$0 < \sum_u \lambda_S^2(u;x_t,s_{t-})h(u;x_t,s_{t-}) \le N\ C_S^2\ C_h.$$

Then,

$$
\begin{aligned}
E([\xi]_t) &= E\int_0^t \xi_{\tau-}^2 d[K]_\tau \\
&= E\left\{\int_0^t \xi_{\tau-}^2 \lambda_D^2(x_\tau, s_\tau)d\tau + \int_0^t \xi_{\tau-}^2 \int_U \lambda_S^2(u; x_\tau, s_{\tau-})m(d\tau, du)\right\} \\
&\leq E\left\{C_D^2 \int_0^t \xi_\tau^2 d\tau + \int_0^t \xi_{\tau-}^2 \int_U \lambda_S^2(u; x_\tau, s_{\tau-})m(d\tau, du)\right\} \\
&\leq E\left\{C_D^2 \int_0^t \xi_\tau^2 d\tau + \int_0^t \xi_{\tau-}^2 \int_U \lambda_S^2(u; x_\tau, s_{\tau-})\gamma_m(d\tau, du)\right\} \\
&\leq E\left\{C_D^2 \int_0^t \xi_{\tau-} d\tau + \int_0^t \xi_{\tau-}^2 \int_U \lambda_S^2(u; x_\tau, s_{\tau-})h(u; x_\tau, s_{\tau-})\eta(du)d\tau\right\} \\
&\leq E\left\{C_D^2 \int_0^t \xi_{\tau-} d\tau + \int_0^t \xi_{\tau-}^2 \sum_u \lambda_S^2(u; x_\tau, s_{\tau-})h(u; x_\tau, s_{\tau-})d\tau\right\} \\
&\leq E\left\{C_D^2 \int_0^t \xi_\tau^2 d\tau + N\, C_S^2\, C_h \int_0^t \xi_{\tau-}^2 d\tau\right\} \\
&\leq C^* \int_0^t E(\xi_\tau^2)d\tau
\end{aligned}
\tag{1.10}
$$

for $C^* = \max(C_D^2, N\, C_S^2\, C_h)$. By the same boundedness and from the direct calculation of expected values under normal and Poissons, we obtain

$$E(\xi_t^2) < C^{**}$$

for some constant C^{**}. This implies $E\big([\xi_t]\big) < \infty$ for all t and hence, ξ_t is a martingale. □

Lemma 2. *Under the risk-neutral probability measure* $\mathbf{Q}$*, the dynamics of state variables,* x_t *and* s_t*, are given by the stochastic differential equations*

$$dx_t = \tilde{\mu}(x_t, s_t)dt + \sigma(x_t, s_t)d\tilde{W}_t \tag{1.11}$$

and

$$ds_t = \int_U (u - s_{t-})\tilde{m}(dt, du) \tag{1.12}$$

where $\tilde{\mu}(x_t, s_t) = \mu(x_t, s_t) - \sigma(x_t, s_{t-})\lambda_D(x_t, s_t)$*;* $\tilde{W}_t$ *is a standard Brownian motion under* $\mathbf{Q}$*;* $\tilde{m}(t, A)$*, the corresponding marked point process of* $m(t, A)$ *under* $\mathbf{Q}$*, has the intensity matrix* $\tilde{H}(u; x_{t-}, s_{t-}) = \{\tilde{h}(u; x_{t-}, s_{t-})\} = \{h(u; x_{t-}, s_{t-})\ (1 - \lambda_S(u; x_{t-}, s_{t-}))\}$*. The compensator of* $\tilde{m}(t, A)$ *under* $\mathbf{Q}$ *becomes*

$$\gamma_{\tilde{m}}(dt, du) = \big(1 - \lambda_S(u; x_{t-}, s_{t-})\big)\gamma_m(dt, du) = \tilde{h}(u; x_{t-}, s_{t-})\eta(du)dt,$$

Proof. Applying Girsanov's theorem on the change of measure for Brownian motion, we have $\tilde{W}_t = W_t - \int_0^t \lambda_D(x_\tau, s_\tau)d\tau$ is a standard Brownian motion under $\mathbf{Q}$. This allows us to obtain $\tilde{\mu}(x_t, s_t) = \mu(x_t, s_t) - \sigma(x_t, s_t)\lambda_D(x_t, s_t)$.

Since the marked point process, $\mu(t, A)$, is actually a collection of $N(N-1)$ conditional Poisson processes, by applying Girsanov's theorem on conditional Poisson process (for example, see Theorems T2 and T3 in Chapter 6 of Bremaud [4]), the conditional Poisson process with intensity, $h(u; x_t, s_{t-})$, under $\mathbf{P}$, becomes the one with intensity, $h(u; x_t, s_{t-})(1 - \lambda_S(u; x_t, s_{t-}))$ under $\mathbf{Q}$. The result follows. □

1.3.4 The term structure of interest rates

Let $\{\mathcal{F}_t\}_{t \geq 0}$ be the natural filtration generated by W and $m(t, \cdot)$. In the absence of arbitrage, the price at time $t-$ of a default-free pure discount bond that matures at T, $P(t-, T)$, can be obtained as,

$$P(t-,T) = E^{\mathbf{Q}}_{t-}\left(e^{-\int_t^T r_\tau d\tau}\right) = E^{\mathbf{Q}}\left\{e^{-\int_t^T r_\tau d\tau}\big|\mathcal{F}_t\right\} = E^{\mathbf{Q}}\left\{e^{-\int_t^T r_\tau d\tau}\big|x_t, s_{t-}\right\} \tag{1.13}$$

with the boundary condition $P(T-, T) = P(T, T) = 1$ and the last equality comes from the Markov property of (x_t, s_t).

Therefore, we can let $P(t-, T) = F(t-, x_t, s_{t-}, T) = F(t-, x, s, T)$ where $x = x_{t-}$ and $s = s_{t-}$. The following proposition gives the partial differential equation characterizing the bond price.

Proposition 1. *The price of the default-free pure discount bond* $F(t-, x, s, T)$ *defined in* (1.13) *satisfies the following partial differential equation*

$$\frac{\partial F}{\partial t} + \tilde{\mu}(x, s)\frac{\partial F}{\partial x} + \frac{1}{2}\sigma^2(x, s)\frac{\partial^2 F}{\partial x^2} + \int_U \Delta_S F\ \tilde{h}(u; x, s)\eta(du) = rF \tag{1.14}$$

with the boundary condition $F(T-, x, s, T) = F(T, x, s, T) = 1$ *and* $\Delta_S F = F(t, x, u, T) - F(t-, x, s, T)$.

Proof. This basically comes from the Feynman-Kac's formula. Or, intuitively, the above result is obtained by applying Ito's formula for semimartingale (Protter [21]) under measure $\mathbf{Q}$ to $F(t, x, s, T)$

$$\begin{aligned} dF =& \Big(\frac{\partial F}{\partial t} + \tilde{u}\frac{\partial F}{\partial x} + \frac{1}{2}\sigma^2\frac{\partial^2 F}{\partial x^2}\Big)dt + \sigma\frac{\partial F}{\partial x}d\tilde{W}_t \\ &+ \big[F(t, x_t, s_t, T) - F(t-, x_{t-}, s_{t-}, T)\big]. \end{aligned} \tag{1.15}$$

Since $x(t)$ is continuous, the last termm in (1.15) can be expressed as

$$\int_U \Delta_S F\ \tilde{m}(dt, du).$$

Note that the above term can be made as a martingale by subtracting its own compensator, which is added back to the dt term. Note that $\gamma_{\tilde{m}}(dt, du) = \tilde{h}(u; x(t-), s(t-))\eta(du)dt$. Therefore $F(t-, x, s, T)$ satisfies the equation

$$\begin{aligned} dF = &\Big\{\frac{\partial F}{\partial t} + \tilde{u}\frac{\partial F}{\partial x} + \frac{1}{2}\sigma^2\frac{\partial^2 F}{\partial x^2} + \int_U \Delta_S F\ \tilde{h}(u;x,s)\eta(du)\Big\}dt \\ &+ \sigma\frac{\partial F}{\partial x}d\tilde{W}_t + \int_U \Delta_s F\ \big[\tilde{m}(dt,du) - \gamma_{\tilde{m}}(dt,du)\big]. \end{aligned} \tag{1.16}$$

Since no arbitrage implies that the instantaneous expected returns of all assets should be equal to the short-term interest rate under the risk-neutral measure, equation (1.14) follows by matching the coefficient of the dt term in (1.16) with that of rF. □

1.4 A tractable specification with exact solution

In general equation (1.14) does not admit a closed-form solution for the bond price. In this section, we consider a tractable specification: an affine term structure of interest rates with regime-switching and regime-switching risk.

1.4.1 Affine regime-switching models

The works of Duffie and Kan [13] and Dai and Singleton [7], among others, provide detailed discussions of completely affine term structure models under diffusions. Duffie, Pan and Singleton [14] deal with general asset pricing models under affine jump-diffusions. Duffee [12] presents a class of *essentially-affine models* and Duarte [11] introduced *semi-affine models*. Both Bansal and Zhou [1] and Landen [19] use affine structure for their regime-switching models. Following this literature, we make the following parametric assumptions

Assumption 1 *The diffusion components of x_t, as well as those in the Markov switching process s_t all have an affine structure. In particular,*

1. $\mu(x_t, s_t) = a_0(s_t) + a_1(s_t)x_t$,
2. $\sigma(x_t, s_t) = \sqrt{\sigma_0(s_t) + \sigma_1 x_t}$,
3. $h(u; x_t, s_{t-}) = \exp\{h_0(u; s_{t-}) + h_1(u; s_{t-})x_t\}$,
4. $\lambda_D(x_t, s_t) = \frac{\lambda_0(s_t)+\lambda_1 x_t + a_1(s_t)x_t}{\sqrt{\sigma_0(s_t)+\sigma_1 x_t}}$,
5. $1 - \lambda_S(u; x_t, s_{t-}) = \frac{e^{\theta(u;s(t-))}}{h(u;x_t,s_{t-})}$.

Assumptions *1* - *3* are related to the two state processes. For the diffusion state process, we assume that the drift term and the volatility term are all affine functions of x_t with regime-dependent coefficients. Then $x(t)$ becomes

$$dx = \big(a_0(s) + a_1(s)\,x\big)\,dt + \sqrt{\sigma_0(s) + \sigma_1\,x}\,dW_t. \tag{1.17}$$

We further assume that the log intensity of regime shifts is an affine function of the short term rate x_t. This assumption ensures the positivity of the intensity function and also allows the transition probability to be time-varying.[5]

Assumptions *4* and *5* deal with the market prices of risk. In the completely affine models, the market price of diffusion risk is proportional to the volatility of the state variable x_t. Such a structure guarantees that the models satisfy a requirement of no-arbitrage, that is, risk compensation goes to zero as risk goes to zero. However, since variances are nonnegative, this structure limits the variation of the compensations that investors anticipate to obtain when encountering a risk. More precisely, since the compensation is bounded below by zero, it cannot change sign over time. This restriction, however, is relaxed in the essentially affine models of Duffee [12].

Following this literature, we also assume that the market price of the diffusion risk is in the form of essentially affine, but we extend to the case with regime-switching with small twists. Specifically, we assume the diffusion risk is a sum of regime-dependent linear combination of x_t and non-regime-dependent scaler of x_t divided by the diffusion coefficient. For the market price of regime-switching risk, we assume a regime-switching dependent constant divided by the intensity of regime switching. We choose these forms of market prices because we may obtain a closed-form solution to the bond prices.

Under these parameterizations of the market prices of risk, the state process x_t and the Markov chain s_t preserve the affine structure. In particular, under the risk-neutral measure $\mathbf{Q}$ the drift term $\tilde{\mu}(s,r)$, and the *log* of regime-switching intensity $\tilde{h}(u;x,s)$ are affine functions of the state x with regime-dependent coefficients. Precisely, under the risk-neutral measure $\mathbf{Q}$,

$$\begin{aligned} dx &= \big(a_0(s) + a_1(s)\,x\big)\,dt + \sqrt{\sigma_0(s) + \sigma_1\,x}\,d\tilde{W}_t - \big[\lambda_0(s) + \lambda_1 x + a_1(s)x\big]\,dt \\ &= \big[a_0(s) - \lambda_0(s) - \lambda_1\,x\big]\,dt + \sqrt{\sigma_0(s) + \sigma_1\,x}\,d\tilde{W}_t. \end{aligned}$$

So, the coefficient

$$\tilde{\mu}(x,s) = a_0(s) - \lambda_0(s) - \lambda_1\,x$$

and $\sigma(x,s)$ remain the same and

$$\tilde{h}(u;x,s(t-)) = e^{\theta(u;s(t-))}.$$

Then, we can solve for the term structure of interest rates and obtain the closed-form solution as follows:

Theorem 2. *Under Assumption 1, the price at time t of a risk-free pure discount bond with maturity τ is given by $f(s(t-),x(t),\tau) = e^{A(\tau,s_t)+B(\tau)x_t}$ and the τ-period interest rate is given by $R(t-,\tau) = -A(\tau,s_{t-})/\tau - B(\tau)x_t/\tau$.*

[5] A more general specification is to allow duration-dependence as well. However a closed-form solution for the yield curve may not be attainable.

With $s = s(t-)$, $A(\tau, s)$ and $B(\tau)$ are determined by the ordinary differential equations

$$-\frac{\partial B(\tau)}{\partial \tau} - \lambda_1 B(\tau) + \frac{1}{2}\sigma_1 B^2(\tau) = \psi_1 \tag{1.18}$$

and

$$\begin{aligned} &-\frac{\partial A(\tau, s)}{\partial \tau} + \big[a_0(s) - \lambda_0(s)\big] B(\tau) + \frac{1}{2}\sigma_0(s) B^2(\tau) \\ &+ \int_U \left[e^{\Delta_S A(\tau,s)} - 1\right] e^{\theta(u;s)} \eta(du) = \psi_0(s) \end{aligned} \tag{1.19}$$

with boundary conditions $A(0, s) = 0$ and $B(0) = 0$, where $\Delta_S A = A(\tau, u) - A(\tau, s)$.

Proof. Without loss of generality, let the price at time $t-$ of a pure discount bond that will mature at T be given by

$$F\big(t-, s(t-), x(t), T\big) = f\big(s(t-), x(t), \tau\big) = e^{A(\tau, s(t-)) + B(\tau)x(t)}$$

where $\tau = T - t$ and $A(0, s) = 0$, $B(0, s) = 0$.

The basic idea is to calculate the derivatives of the above bond price F, substitute them in equation(1.14), and match the coefficients of x.

Observe that

$$\frac{\partial F}{\partial \tau} = F\Big(-\frac{\partial A(\tau, s)}{\partial \tau} - \frac{\partial B(\tau)}{\partial \tau}x\Big), \quad \frac{\partial F}{\partial x} = FB(\tau), \quad \frac{\partial^2 F}{\partial x^2} = FB^2(\tau),$$

$$\tilde{h}\big(u; x(t), s(t-)\big) = h\big(u; x(t), s(t-)\big)\big(1 - \lambda_S(u; x(t), s(t-))\big) = e^{\theta(u; s(t-))},$$

$$F_S = F(e^{\Delta_S A} - 1)$$

where $\Delta_S A = A(\tau, u) - A(\tau, s)$, and recall that

$$r = \psi_0(s) + \psi_1 x.$$

With some simplifications and letting $s = s(t-)$, Proposition 1 then implies

$$\begin{aligned} \psi_0(s) + \psi_1 x = &-\frac{\partial A(\tau, s)}{\partial \tau} - \frac{\partial B(\tau)}{\partial \tau}x + \big[a_0(s) - \lambda_0(s) - \lambda_1 x\big] B(\tau) \\ &+ \frac{1}{2}\big[\sigma_0(s) + \sigma_1 x\big] B^2(\tau) + \int_U \left(e^{\Delta_S A} - 1\right) e^{\theta(u;s)} \eta(du) \end{aligned} \tag{1.20}$$

Then, Theorem 2 follows by matching the coefficients of x on both sides of the above equation. □

The above model extends the existing literature on the term structure of interest rates under regime shifts in several ways. While Landen [19] provided an exact solution to the yield curve only under risk-neutral probability measure, there was no mention of the market price of regime-switching risk. The survey paper of Dai and Singleton [8] proposed a Gaussian affine-type model with regime-switching risk and constant volatility within each regime. In our model, we allow for stochastic volatilities in each regime and our diffusion risk is in an essentially-affine form. In the case of Bansal and Zhou [1], the risk of regime shifts is not priced either, and they had to rely on log-linear approximation to obtain closed-form solution for bond pricing.

Finally, we examine the expected excess return on a long term bond over the short rate implied by our model.

Corollary 1. *Under the assumptions of Theorem 2, the expected excess return on a long term bond over the short rate is given by*

$$E_t\left(\frac{df_t}{f_{t-}}\right) - r_t dt = \big[\lambda_0(s) + \lambda_1 x + a_1(s)x\big]B(\tau)dt + \int_U \left(e^{\Delta_S A} - 1\right)\left(e^{h_0(u;s)+h_1(u;s)x} - e^{\theta(u;s)}\right)\eta(du)dt. \tag{1.21}$$

Proof. Similary, let the price at time $t-$ of a pure discount bond that will mature at T be given by

$$F(t-, s(t-), x(t), T) = f\big(s(t-), x(t), \tau\big) = e^{A(\tau, s(t-)) + B(\tau)x(t)}$$

where $\tau = T - t$ and $A(0, s) = 0$, $B(0, s) = 0$.

Applying Itô's formula to $F(t-, s(t-), x(t), T)$ under the physical measure $\mathbf{P}$, we obtain the following equation similar to (1.16):

$$dF = \left\{\frac{\partial F}{\partial t} + u\frac{\partial F}{\partial x} + \frac{1}{2}\sigma^2\frac{\partial^2 F}{\partial x^2} + + \int_U \Delta_S F\ h(u; x, s)\eta(du)\right\}dt + \sigma\frac{\partial F}{\partial x}dW_t + \int_U \Delta_s F\ \big[m(dt, du) - \gamma_m(dt, du)\big]. \tag{1.22}$$

Applying Proposition 1, we wish to make the coefficient of the dt term equal to rF by subtracting and adding terms. Using Lemma 2, noting the last two terms are martingales, and taking conditional expectation with some simplifications, we obtain

$$E_t\left(\frac{dF}{F}\right) - r_t dt = \sigma(x, s)\lambda_D(x, s)\frac{\partial F}{\partial x}/F dt + \int_U \frac{\Delta_S F}{F} h(u; x, s)\ \lambda_S(u; x, s)\eta(du)dt. \tag{1.23}$$

With simplifications, Assumption 1 implies that the above equation becomes equation (1.21). □

The first term on the right hand side of equation (1.21) is interpreted as the diffusion risk premium in the literature, and the second term can be analogously defined as the regime-switching risk premium. The equation shows that introducing the dependence of the market prices of diffusion on s_t adds more flexibility to the specification of the risk premium. Bansal and Zhou [1] points out that it is mainly this feature of the regime-switching model that provides improved goodness-of-fit over the existing term structure models. On the other hand, (1.21) also shows that if the term structure exhibits significant difference across regimes ($\Delta_s A \neq 0$), there is an additional source of risk due to regime shifts and it should also be priced ($e^{h_0(u;s)+h_1(u;s)x} - e^{\theta(u;s)}$) in the term structure model. Introducing the regime switching risk not only can add more flexibilities to the specification of time-varying bond risk premiums, but also can be potentially important in understanding the bond risk premia over different holding periods.

1.5 Conclusions

This paper first presents a new marked point process representation of regime change using a random measure. We apply this new representation to specify a term structure model of interest rate with regime-switching risk. We derive an exact solution for the yield curve in an essentially-affine specification.

With this exact solution, we can further estimate the model by the efficient method of moments as in Wu and Zeng [22] and quantify the regime-switching risk and its impact on yield curves. Other important topics that can be explored further include the implications and impacts of regime-switching risk on bond derivatives, and on investors' optimal portfolio choice problem. Also, more studies are needed on the empirical evidence of regime-switching risk in interest rates. These topics are left for future research.

References

1. Bansal, R. and H. Zhou (2002). "Term structure of interest rates with regime shifts". *Journal of Finance*, 57: 1997-2043.
2. Bielecki, T. and M. Rutkowski (2000). "Multiple ratings model of defaultable term structure". *Mathematical Finance*, 10: 125-139.
3. Bielecki, T. and M. Rutkowski (2001). "Modeling of the defaultable term structure: conditional Markov approach". Working Paper, The Northeastern Illinois University.
4. Bremaud, P. (1981). *Point Processes and Queues, Martingale Dynamics.* Springer-Verlag, Berlin.
5. Cox, J., J. Ingersoll and S. Ross (1985a). "An intertemporal general equilibrium model of asset prices". *Econometrica* 53: 363-384.
6. Cox, J., J. Ingersoll and S. Ross (1985b). "A theory of the term structure of interest rates". *Econometrica* 53: 385-407.

7. Dai, Q. and K. Singleton (2000). "Specification analysis of affine term structure models". *Journal of Finance*, 55: 1943-78.
8. Dai, Q. and K. Singleton (2003). "Term structure dynamics in theory and reality". *Review of Financial Studies* 16: 631-678.
9. Dai, Q., K. Singleton and W. Yang (2006). "Regime shifts in a dynamic term structure model of the u.s. treasury bond yields". Working Paper, Stanford University.
10. Diebold, F. and G. Rudebusch (1996). "Measuring business cycles: A modern perspective". *Review of Economics and Statistics* 78: 67-77.
11. Duarte, J. "Evaluating an alternative risk preference in affine term structure models". *Review of Financial Studies*, 17: 379 - 404.
12. Duffee, G. (2002). "Term premia and interest rate forecasts in affine models". *Journal of Finance*, 57: 405-443.
13. Duffie, D. and R. Kan (1996). "A yield-factor model of interest rates". *Mathematical Finance* 6: 379-406.
14. Duffie, D., J, Pan and K. Singleton (2000). "Transform analysis and asset pricing for affine jump-diffusions". *Econometric* 68: 1343-1376.
15. Elliott, R. J. et. al. (1995). *Hidden Markov Models: Estimation and Control.* New York, Springer-Verlag.
16. Elliott, R. J. and R. S. Mamon (2001). "A complete yield curve descriptions of a Markov interest rate model", *International Journal of Theoretical and Applied Finance*, 6: 317-326.
17. Hamilton, J. (1989). "A new approach to the economic analysis of nonstationary time series and the busines cycle". *Econometrica* 57: 357-384.
18. Harrison, M. and D. Kreps (1979). "Martingales and arbitrage in multiperiod security markets". *Journal of Economic Theory*, 20: 381-408.
19. Landen, C. (2000). "Bond pricing in a hidden Markov model of the short rate". *Finance and Stochastics* 4: 371-389.
20. Last, G. and A. Brandt (1995). *Marked Point Processes on the Real Line.* Springer, New York.
21. Protter, P. (2003). *Stochastic Intergration and Differential Equations.* 2nd edition, Springer, Berlin.
22. Wu, S. and Y. Zeng (2005). "A general equilibrium model of the term structure of interest rates under regime-switching risk", *International Journal of Theortical and Applied Finance*, 8: 839-869.
23. Wu, S. and Y. Zeng (2006). "The term structure of interest rates under regime shifts and jumps". *Economics Letters*, (in press).
24. Yin, G.G. and Q. Zhang (1998). *Continuous-time Markov Chains and Applications. A Singular Perturbation Approach..* Berlin, Springer.

2

The Term Structure of Interest Rates in a Hidden Markov Setting

Robert J. Elliott[1] and Craig A. Wilson[2]

[1] Haskayne School of Business
University of Calgary
Calgary, Alberta, Canada
relliott@ucalgary.ca

[2] College of Commerce
University of Saskatchewan
Saskatoon, Saskatchewan, Canada
cwilson@commerce.usask.ca

Summary. We describe an interest rate model in which randomness in the short-term interest rate is partially due to a Markov chain. We model randomness through the volatility and mean-reverting level as well as through the interest rate directly. The short- term interest rate is modeled in a risk-neutral setting as a continuous process in continuous time. This allows the valuation of interest rate derivatives using the martingale approach. In particular, a solution is found for the value of a zero-coupon bond. This leads to a non-linear regression model for the yield to maturity, which is used to filter the state of the unobservable Markov chain.

Key words: Interest rate modeling, term structure, filtering, Markov chain

2.1 Introduction

Current models of the short-term interest rate often involve treating the short rate as a diffusion or jump diffusion process in which the drift term involves exponential decay toward some value. The basic models of this type are Vasiček [10] and Cox, Ingersoll and Ross [2], where the distinction between these two interest rate models rests with the diffusion term. The drift term, (of both models), tends to cause the short rate process to decay exponentially towards a *constant* level. This feature is responsible for the mean-reverting property exhibited by these processes.

An extension to these models has come in the form of allowing the drift to incorporate exponential decay toward a *manifold*, rather than a constant. This is known as the Hull and White [7] model, and it allows the short rate process the tendency to follow the initial term structure of interest rates. This is an important extension, because with a judicious choice of the manifold, the initial term structure predicted by the model can exactly match the existing term structure, and because of this feature, models of this class are called no arbitrage models. In general, this cannot be done with a constant mean-reverting level, and such models are often called equilibrium models, since they generate stationary interest rate processes. Although there are many other extensions to the basic models—incorporating stochastic volatility, non-linear drift (so decay is no longer exponential), and jumps, for example—the Hull-White extension is the most applicable to the bond pricing component of our study.

The Hull-White model has many advantages: it possesses a closed-form solution for the price of zero-coupon bonds, as well as for call options on such bonds, and it can also be calibrated to fit the initial yield curve exactly. However, one of the disadvantages of the model is that, because there is only one factor of randomness, it only allows parallel shifts in the yield curve through time. Bonds of all maturities are necessarily perfectly correlated with each other. This approach cannot explain the common phenomenon of yield curve twists. This motivates the need to incorporate an additional factor of randomness into the basic model.

The Hull-White model is described under the risk-neutral probability by the stochastic differential equation

$$dr_t = a(t)\{\bar{r}(t) - r_t\}\, dt + \sigma(t) r_t^{\rho} dw_t,$$

where r_t represents the short-term, continuously compounded interest rate, and $\{w_t\}$ is a Brownian motion under the risk-neutral probability. The parameter ρ takes one of the two values 0 or $1/2$, depending on whether it extends the Vasiček or Cox-Ingersoll-Ross model. The parameter functions $a(t)$, $\bar{r}(t)$, and $\sigma(t)$ extend the basic models, in which these parameters are just constants. The randomness in this model comes from the Brownian motion, and for the extended Vasiček model when $\rho = 0$, it can be interpreted as adding white noise to the short rate. For the extended Cox-Ingersoll-Ross model the noise is multiplicative, but it is still applied directly to the short rate process.

The main problem with this model is in the way it handles the cyclical nature of interest rates. A time series of interest rates tends to appear cyclical because the supply and demand for money is closely related to income growth, which fluctuates with the business cycle. This has implications for real (adjusted for inflation) interest rates. For example, at a business cycle peak short-term rates should be rising and at a trough rates should be falling. This also has implications for the slope of the term structure—it should be steeper at a peak and flatter at a trough. Roma and Torous [9] find that this property of real interest rates cannot be explained by a simple additive noise type model, such as Vasiček. The Hull-White extension can provide a correction for this problem to a degree, but since the parameter functions are deterministic, it implies that the business cycle effects are known with certainty, which does not allow for the possible variation in length and intensity from what is expected. In addition, when the central bank targets a constant rate of inflation, this fluctuation is transferred to nominal interest rates, so the same characteristics could apply to them.

We approach this problem by modeling the mean-reverting level directly as a random process, and have the short rate chase the mean-reverting level in a linear drift type model. This is similar to the model proposed by Balduzzi, Das, and Foresi [1], except instead of a diffusion process, here the mean-reverting level is assumed to follow a finite-state, continuous-time Markov chain. The switching of the Markov chain to different levels produces a cyclical pattern in the short rate that is consistent with the above effect, and the randomness inherent in the Markov chain prevents the business cycle lengths and intensities from being completely predictable.

The remainder of this paper is organized as follows. Section 2.2 discusses the model, including details about the Markov chain, the short-term interest rate, and the term-structure model. Section 2.3 outlines how the model is implemented and Section 2.4 provides the results of implementing it and discusses some implications. Finally Section 2.5 concludes.

2.2 The Model

In this section we construct the model of the short-term interest rate. This model will be used to derive prices for bonds.

We begin by describing the probability space, denoted by $(\Omega, \mathcal{F}, P)$, that is used to model randomness in this framework. We assume that P is a risk-neutral probability measure, whose existence can be assured by an absence of arbitrage in the underlying economy. Furthermore, we assume that the σ-field over Ω, $\mathcal{F}$, is complete and large enough to support the increasing filtration of sub-σ-fields $\{\mathcal{F}_t\}$ associated with the Markov chain and Brownian motion described below.

2.2.1 The Markov chain

A stochastic process, $\{X_t\}$ satisfies the Markov property (with respect to probability P and filtration $\{\mathcal{F}_t\}$) if

$$P\{X_{s+t} \in B|\mathcal{F}_s\} = P\{X_{s+t} \in B|X_s\}$$

for all $s, t \geq 0$ and all Borel sets, B. If such a stochastic process takes values in a countable set, it is called a Markov chain.

For our purposes, we consider a Markov chain generated by a transition rate matrix $\mathbf{Q}$. Here $\mathbf{Q}$ is an $N \times N$ conservative Q-matrix with non-negative off-diagonal entries and rows that sum to zero. In general, $\mathbf{Q}$ could change with time, but for simplicity and without any direction about how it should change we assume that $\mathbf{Q}$ is constant or homogeneous in time.

A transition function for a Markov chain relates the probability of changing from one state to another within a certain time, and the transition matrix is constructed so that each entry is a transition function

$$\mathbf{P}_{ij}(s,t) = P\{X_{s+t} = j | X_s = i\}.$$

The transition matrix $\mathbf{P}$ for a Markov chain can be generated by the transition rate matrix $\mathbf{Q}$ through the forward Kolmogorov equation

$$\frac{\partial \mathbf{P}(s,t)}{\partial t} = \mathbf{P}(s,t)\mathbf{Q}(s+t).$$

Since $\mathbf{Q}$ is homogeneous, the general solution to the forward Kolmogorov equation is $\mathbf{P}(s,t) = \mathbf{C}(s)e^{\mathbf{Q}t}$ and since $\mathbf{P}(s,0) = \mathbf{I}$, the identity matrix, the constant must also be the identity matrix, $\mathbf{C}(s) = \mathbf{I}$. From this we can conclude that the transition functions are independent of the starting time s, the transition matrix is the matrix exponential of $\mathbf{Q}$

$$\mathbf{P}(t) = e^{\mathbf{Q}t},$$

and the forward Kolmogorov integral equation is

$$\mathbf{P}(t) = \mathbf{I} + \int_0^t \mathbf{P}(u)\mathbf{Q}\, du.$$

Without loss of generality, we assume that the Markov chain is right continuous and it takes values from the set of canonical unit vectors of R^N, $\{\mathbf{e}_1, ..., \mathbf{e}_N\}$, where $\mathbf{e}_i$ is the vector with 1 in the ith entry and 0 elsewhere. To make this clear we will denote the Markov chain by $\{\mathbf{x}_t\}$. In this case we have $E[\mathbf{x}_t] = \mathbf{P}(t)^\top \mathbf{x}_0$, which is the probability distribution for the Markov chain and where $^\top$ denotes the transpose of a vector. More generally we have

$$E[\mathbf{x}_{s+t}|\mathbf{x}_s] = \mathbf{P}(t)^\top \mathbf{x}_s.$$

Putting this together with the forward Kolmogorov equation gives

$$E[\mathbf{x}_{s+t}|\mathbf{x}_s] = \mathbf{x}_s + \int_0^t \mathbf{Q}^\top \mathbf{P}(u)^\top \mathbf{x}_s\, du.$$

It follows directly from this that the stochastic process

$$\mathbf{m}_t = \mathbf{x}_t - \mathbf{x}_0 - \int_0^t \mathbf{Q}^\top \mathbf{x}_u\, du$$

is a square-integrable, right-continuous, zero-mean martingale. Therefore, $\{\mathbf{x}_t\}$ is a semi-martingale

$$\mathbf{x}_t = \mathbf{x}_0 + \int_0^t \mathbf{Q}^\top \mathbf{x}_u \, du + \mathbf{m}_t.$$

(This derivation is adapted from Elliott [4].)

There are a number of other benefits that arise from associating states of the Markov chain with unit vectors. First note that the inner product of the Markov chain at any time is always equal to 1

$$\mathbf{x}_t^\top \mathbf{x}_t = 1,$$

the inner product between $\mathbf{1}$, the vector with 1 in each entry, and the Markov chain is also always 1

$$\mathbf{1}^\top \mathbf{x}_t = 1,$$

and the outer product of the Markov chain is the diagonal matrix of the Markov chain

$$\mathbf{x}_t \mathbf{x}_t^\top = \mathrm{diag}[\mathbf{x}_t].$$

Furthermore, any real-valued function of the Markov chain has a linear representation

$$f(\mathbf{x_t}) = \mathbf{f}^\top \mathbf{x}_t$$

where $\mathbf{f}_i = f(\mathbf{e}_i)$ and any vector-valued function of the Markov chain also has a linear representation

$$\mathbf{f}(\mathbf{x_t}) = \mathbf{F}^\top \mathbf{x}_t$$

where $\mathbf{F}_{ij} = \mathbf{f}_j(\mathbf{e}_i)$. Finally, notice that iterated multiples of the Markov chain have the following idempotency property

$$(\mathbf{f}^\top \mathbf{x}_t)\mathbf{x}_t = \mathrm{diag}[\mathbf{f}]\mathbf{x}_t.$$

The linear representations are also useful for describing the dynamics of certain stochastic processes. Consider the stochastic process $\{\mathbf{f}_t\}$ where $\mathbf{f}_t = \mathbf{F}_t^\top \mathbf{x}_t$, and $\mathbf{F}_t$ is continuous and adapted to $\{\mathcal{F}_t\}$. Then applying Itô's integration by parts for general semi-martingales allows the semi-martingale decomposition

$$\mathbf{f}_t = \mathbf{f}_0 + \int_0^t \mathbf{F}_u^\top \mathbf{Q}^\top \mathbf{x}_u \, du + \int_0^t \{d\mathbf{F}_u^\top \mathbf{x}_u\} + \int_0^t \mathbf{F}_u^\top \, d\mathbf{m}_u.$$

For the special case where $\mathbf{F}_u^\top$ commutes with $\mathbf{Q}^\top$ and $d\mathbf{F}_u^\top = \mathbf{G}_u^\top \mathbf{F}_u^\top \, du$, the semi-martingale representation can be written as

$$\mathbf{f}_t = \mathbf{f}_0 + \int_0^t \{\mathbf{Q} + \mathbf{G}_u\}^\top \mathbf{f}_u \, du + \int_0^t \mathbf{F}_u^\top \, d\mathbf{m}_u.$$

The following particular example arises in the context of bond pricing. Consider the processes $f_t = \exp(\int_0^t \mathbf{g}_u^\mathsf{T}\mathbf{x}_u\, du)$, where $\{\mathbf{g}_t\}$ is adapted, and $\mathbf{F}_t = f_t\mathbf{I}$. Clearly $\{\mathbf{F}_t\}$ is continuous, adapted, and it commutes with any $N \times N$ matrix. Furthermore, $d\mathbf{F}_t = \mathbf{F}_t\mathbf{G}_t\, dt$, where $\mathbf{G}_t = \mathbf{g}_t^\mathsf{T}\mathbf{x}_t\mathbf{I}$, so the dynamics of $\mathbf{f}_t = \mathbf{F}_t^\mathsf{T}\mathbf{x}_t$ have the above semi-martingale form. Moreover, $\mathbf{G}_t^\mathsf{T}\mathbf{f}_t = \mathrm{diag}[\mathbf{g}_t]\mathbf{f}_t$, so we have the semi-martingale representation

$$\mathbf{f}_t = \mathbf{f}_0 + \int_0^t \{\mathbf{Q} + \mathrm{diag}[\mathbf{g}_u]\}^\mathsf{T}\mathbf{f}_u\, du + \int_0^t \mathbf{F}_u^\mathsf{T}\, d\mathbf{m}_u.$$

If $\mathbf{g}_t$ and $\mathbf{f}_t$ are independent, (for example if $\mathbf{g}_t$ is deterministic), then we can find $E[\mathbf{f}_t]$ by solving a homogeneous linear system of ordinary differential equations. Since we can equivalently write $\mathbf{f}_t = f_t\mathbf{x}_t$, we have $f_t = \mathbf{1}^\mathsf{T}\mathbf{f}_t$, and therefore $E[f_t] = \mathbf{1}^\mathsf{T}E[\mathbf{f}_t]$.

2.2.2 The short-term interest rate

We now consider the model for the short-term interest rate. The short rate dynamics are defined through a stochastic differential equation, so a priori we require that a Brownian motion denoted $\{w_t\}$ exists for our probability space and filtration. In fact, since Brownian motion is a martingale, it is straight forward to show by taking $\mathbf{F}_t = w_t\mathbf{I}$ above that if it exists, it must be uncorrelated with the Markov chain. However, we require that the Markov chain and Brownian motion be independent, so we will assume that this stronger condition is satisfied. In this case, by defining $\{\mathcal{F}_t^x\}$ to be the filtration generated by the Markov chain, $\{w_t\}$ is still a Brownian motion with respect to the larger filtration $\{\mathcal{F}_t \vee \mathcal{F}_T^x\}$ for fixed T.

Following Naik and Lee [8], we model the short rate dynamics denoted $\{r_t\}$ using the equation

$$dr_t = a(\bar{r}_t - r_t)\, dt + \sigma_t\, dw_t. \tag{2.1}$$

This model suggests that the short rate is expected to decay exponentially toward the level $\bar{r}_t$ at the rate a, but it is subjected to additive noise modulated by the volatility σ_t. The level and volatility parameters are permitted to switch from time to time according to the state of the Markov chain, so we have

$$\bar{r}_t = \bar{\mathbf{r}}^\mathsf{T}\mathbf{x}_t \quad \text{and} \quad \sigma_t = \sigma^\mathsf{T}\mathbf{x}_t.$$

For simplicity and estimation purposes, we take the parameters to be constant, but the analysis follows identically if these are functions of time. This specification has two benefits over the basic models of Vasiček [10] and Hull and White [7]. It allows a better fit to the term structure and it has the potential to resolve the difficulty with accurately estimating the mean r eversion rate a.

The solution to the SDE in (2.1) is

$$r_t = \frac{1}{A_t}\left\{r_0 + \int_0^t A_u a\bar{r}_u\, du + \int_0^t A_u \sigma_u\, dw_u\right\} \tag{2.2}$$

where

$$A_t = e^{\int_0^t a\, du}.$$

The more general version of (2.2) is

$$r_t = \frac{1}{A_t}\left\{A_s r_s + \int_s^t A_u a\bar{r}_u\, du + \int_s^t A_u \sigma_u\, dw_u\right\} \quad \text{for } s \le t$$

From this, we can see that conditional on the information $\mathcal{F}_t^x$, r_t is normally distributed. Furthermore, by changing the order of integration we have

$$\int_0^t r_u\, du = r_0 \int_0^t \frac{A_0}{A_s}\, ds + \int_0^t \left(\int_u^t \frac{A_u}{A_s}\, ds\right) a\bar{r}_u\, du + \int_0^t \left(\int_u^t \frac{A_u}{A_s}\, ds\right) \sigma_u\, dw_u.$$

Again, conditional on $\mathcal{F}_t^x$, $\int_0^t r_u\, du$ has a normal distribution with mean and variance

$$E\left[\int_0^t r_u\, du \,\middle|\, \mathcal{F}_t^x\right] = r_0 \int_0^t \frac{A_0}{A_s}\, ds + \int_0^t \left(\int_u^t \frac{A_u}{A_s}\, ds\right) a\bar{r}_u\, du$$

$$\text{var}\left[\int_0^t r_u\, du \,\middle|\, \mathcal{F}_t^x\right] = \int_0^t \left(\int_u^t \frac{A_u}{A_s}\, ds\right)^2 \sigma_u^2\, du.$$

2.2.3 The zero-coupon bond value

Since we are working under the risk-neutral probability, the value of a zero-coupon bond maturing in t years is

$$B(t) = E\left[\exp\left(-\int_0^t r_u\, du\right)\right].$$

We determine this expectation in two stages, by first conditioning on the σ-field $\mathcal{F}_t^x$. Because the integral is conditionally normal, it is straightforward to get the conditional expectation

$$\begin{aligned} E\left[\exp\left(-\int_0^t r_u\, du\right) \,\middle|\, \mathcal{F}_t^x\right] &= \exp\left\{\frac{1}{2}\int_0^t \left(\int_u^t \frac{A_u}{A_s}\, ds\right)^2 \sigma_u^2\, du \right. \\ &\qquad \left. - r_0 \int_0^t \frac{A_0}{A_s}\, ds - \int_0^t \left(\int_u^t \frac{A_u}{A_s}\, ds\right) a\bar{r}_u\, du\right\} \\ = \exp\left(-r_0 \int_0^t \frac{A_0}{A_s}\, ds\right) &\exp\left\{\int_0^t \left\{\frac{1}{2}\left(\int_u^t \frac{A_u}{A_s}\, ds\right)^2 \sigma^2 - \left(\int_u^t \frac{A_u}{A_s}\, ds\right) a\bar{\mathbf{r}}\right\}^{\mathsf{T}} \mathbf{x}_u\, du\right\} \end{aligned} \tag{2.3}$$

where the first term in equation (2.3) is deterministic. Therefore, we find the zero-coupon bond price by taking the expected value of the second term.

This is similar to the situation described at the end of Subsection 2.2.1. However, in this case the integrand is also a function of t. To deal with this, we fix a maturity time T and define a function

$$g_u = \frac{1}{2}\left(\int_u^T \frac{A_u}{A_s}\,ds\right)^2 \sigma^2 - \left(\int_u^T \frac{A_u}{A_s}\,ds\right) a\bar{\mathbf{r}}.$$

This quantity is deterministic and with a constant rate of mean reversion a we get

$$\int_u^T \frac{A_u}{A_s}\,ds = \frac{1-e^{-a(T-u)}}{a}.$$

Carrying on with the previous notation f_t and $\mathbf{f}_t$, we find the expectation $E[\mathbf{f}_t]$ by solving the homogeneous linear ordinary differential equation

$$\mathbf{y}'(t) = \left\{\mathbf{Q} + \mathrm{diag}[\mathbf{g}_t]\right\}^{\mathsf{T}} \mathbf{y}(t). \tag{2.4}$$

Calling the fundamental matrix in equation (2.4) $\mathbf{\Phi}(t)$ and noting that the initial value is $\mathbf{f}_0 = \mathbf{x}_0$, we write $E[\mathbf{f}_t] = \mathbf{\Phi}(t)\mathbf{x}_0$ and $E[f_t] = \mathbf{1}^{\mathsf{T}}\mathbf{\Phi}(t)\mathbf{x}_0$. Evaluating this at $t = T$ gives the value of a zero-coupon bond maturing at time T

$$B(T) = \exp\left(-r_0 \int_0^T \frac{A_0}{A_s}\,ds\right) \mathbf{1}^{\mathsf{T}}\mathbf{\Phi}(T)\mathbf{x}_0.$$

This is fine when the Markov chain is observable, but in our case we consider the Markov chain hidden. This means that the above bond value is still based on a conditional expectation given $\mathcal{F}_0^x$, and taking expected value requires replacing $\mathbf{x}_0$ with $E[\mathbf{x}_0] = \bar{\mathbf{x}}_0$. In other words, because the Markov chain is hidden, we must base our decisions on the probability distribution of its states. The continuously compounded yield to maturity of such a bond is

$$R(T) = \frac{r_0}{T}\int_0^T \frac{A_0}{A_s}\,ds - \frac{\ln\left(\mathbf{1}^{\mathsf{T}}\mathbf{\Phi}(T)\bar{\mathbf{x}}_0\right)}{T}. \tag{2.5}$$

2.3 Implementation

We implement this model using 7 years of monthly US term structure data from January 1999 to December 2005. The dataset was obtained from the Fama risk-free rate and Fama-Bliss discount structure files of the CRSP database. This data provides continuously compounded yield to maturity on 1-month, 3-month, 6-month, and 1-year US T-bills, and it constructs continuously compounded yield to maturity on hypothetical zero-coupon US treasury bonds with maturities ranging annually from 2 to 5 years. This gives eight different maturities observed over 84 months for a total of 672 observations. A quick scan of the data revealed that the July 2003 observation of the six-month yield was erroneously recorded as zero, so we drop this observation leaving a total of 671 remaining observations.

The theoretical yield to maturity (2.5) derived in Subsection 2.2.2 provides a natural non-linear regression model to apply to this data. Writing

$$\alpha(T) = \frac{1}{T}\int_0^T \frac{A_0}{A_s}\,ds = \frac{1-e^{-aT}}{aT}$$

and $R_t(T)$ for the theoretical yield to maturity on a zero-coupon bond at time t that matures T years from then at time $t+T$ we get

$$R_t(T) = \alpha(T) r_t - \frac{\ln\left\{\mathbf{1}^{\mathsf{T}}\boldsymbol{\Phi}(T)\bar{\mathbf{x}}_t\right\}}{T}.$$

It is tempting to formulate the second term as a linear function of $\bar{\mathbf{x}}_t$; however, we cannot do this since $\bar{\mathbf{x}}_t$ is not a unit vector as $\mathbf{x}_t$ is. Denoting the observed data as $y_{t,T}$ where T represents the maturity and t represents the date, leads to the regression equation

$$y_{t,T} = R_t(T) + \epsilon_{t,T}.$$

We assume that the residuals $\{\epsilon_{t,T}\}$ are independent with mean 0 and variance η^2. This approach involves minimizing the sum of squared errors or residuals between the predicted theoretical yield and the actual yield observed in the data. Since the theoretical yield is not dynamic in the sense that it does not depend on lagged observations, the parameter estimators are weakly consistent provided the residuals have finite variance $\eta^2 < \infty$, which we will assume to be the case. For details on this see Davidson and MacKinnon [3].

There are three main difficulties we face in implementing the model using non-linear regression. First, in order to solve the differential equation we need the dimension of the Markov chain's state space. Expanding the state space can only reduce the sum of squared residuals because a model with a smaller state space can be considered a nested restriction of a more general model. The restriction could come in the form of requiring both mean-reverting level and volatility values to be equal in two particular states. Because of this, it is impossible to use non-linear regression to estimate the proper dimension of the Markov chain state space. To find an appropriate dimension, an F test could be used to determine when the improvement from increasing the dimension is no longer significant. Therefore, we need to fix the dimension of the Markov chain's state space before running the regression.

The next difficulty involves solving the differential equation. Since the coefficient matrix depends on time t, the fundamental solution matrix does not have a well-known closed form such as an exponential matrix. Therefore we solve the differential equation numerically. We do this by approximating the differential equation with the following difference equation

$$\boldsymbol{\Phi}\big((n+1)\Delta t\big) = \left\{\mathbf{I} + \big(\mathbf{Q} + \operatorname{diag}[\mathbf{g}_{n\Delta t}]\big)\Delta t\right\}^{\mathsf{T}}\boldsymbol{\Phi}(n\Delta t).$$

The solution uses $n = 1000$ intervals for each maturity, so $\Delta t = T/1000$. This provides a degree of accuracy of at least five significant digits for each element of the fundamental matrix $\mathbf{\Phi}(T)$ for all maturities up to $T = 5$ years.

The final problem we face deals with the initial values for the short-term interest rate and the Markov chain, r_t and $\bar{\mathbf{x}}_t$. Neither of these is provided by the data source CRSP. Since the Markov chain is unobservable, there is no hope of finding data elsewhere to use for its initial value at any date. Therefore we must estimate the initial probability distributions for the Markov chain at each date. This is a classic filtering problem and one way to approach it is to use a discrete version of the short rate dynamics

$$\Delta r_t = a(\bar{\mathbf{r}}^{\mathsf{T}}\mathbf{x}_t - r_t)\Delta t + \sigma^{\mathsf{T}}\mathbf{x}_t \Delta w_t$$

and monthly observations of the short-term interest rate for the desired period January 1991 to December 2005. An extension of the filtering techniques described in Elliott [5] can be applied to such a problem to get maximum likelihood estimates of the Markov chain state probabilities. Unfortunately, this also requires observation of the short-term interest rate, but more importantly it requires that the Markov chain transition probabilities be the same under the true measure, which is used by the filtering procedure and the risk-neutral measure, which is needed for the term-structure model.

A simpler filtering approach can be devised for our situation. We can simply treat the initial Markov chain state probabilities at each date as unknown parameters in our non-linear regression. Then the parameter estimates produce a filter for the state of the Markov chain and this automatically ensures that from the perspective of minimizing the sum of squared errors for the series of term structures the optimal filter is used. Unfortunately, since the Markov chain state probabilities do not enter the regression equation linearly, this optimal filter cannot be expressed analytically, so the values must be obtained numerically.

Turning our attention back to the initial short-term interest rate at each date, we again have two alternatives. We can use a proxy for the short rate such as the Federal Funds overnight rate, or we can filter values for the initial short rate at each date using our non-linear regression model and the term structure data. Naturally this latter approach uses up many more degrees of freedom by requiring 180 additional estimates. On the other hand, the filtering approach will choose these values optimally. Since the initial interest rate does enter the theoretical yield formula linearly, the optimal value is found to be

$$r_t^* = \frac{\sum_T \left\{ y_{t,T} + \frac{\ln\{\mathbf{1}^{\mathsf{T}}\mathbf{\Phi}(T)\bar{\mathbf{x}}_t\}}{T} \right\} \alpha(T)}{\sum_T \alpha(T)^2}.$$

These optimal values may differ substantially from the proxy values. One reason for this difference may have to do with the institutional features of the

US banking system that increase demand and thus price for treasury securities beyond an optimal competitive level. In any case, an F test can be used to determine whether the model is significantly hindered by considering the more parsimonious restricted model with the initial short rate proxied by the Federal Funds overnight rate. Next we look at the results from implementing the model.

2.4 Results

In this section we present the results from implementing the term structure model in several situations. Table 2.1 provides the main parameter estimates

	$N=1$			$N=3$	
	Fed Fund	Free Est.		Fed Fund	Free Est.
a	0.238872	0.203953	a	0.628730	0.670303
$\bar{r}$	0.059517	0.068345	$\bar{\mathbf{r}}_1$	0.669713	0.355563
σ	0.000114	0.000114	$\bar{\mathbf{r}}_2$	–0.07240	–0.10679
std err	0.004691	0.003095	$\bar{\mathbf{r}}_3$	–0.71369	–0.12026
F_{fed}		11.31533	σ_1	0.024351	0.023138
	$N=2$		σ_2	0.024041	0.023241
a	0.404457	0.575902	σ_3	0.000187	0.000187
$\bar{\mathbf{r}}_1$	0.275108	0.105218	$\mathbf{Q}_{12}$	0.040939	0.030791
$\bar{\mathbf{r}}_2$	–0.24453	–0.01730	$\mathbf{Q}_{13}$	5.723191	1.423016
σ_1	0.000185	0.000185	$\mathbf{Q}_{21}$	0.181682	0.229825
σ_2	0.000611	0.000611	$\mathbf{Q}_{23}$	0.287840	0.207573
$\mathbf{Q}_{12}$	0.307727	0.214106	$\mathbf{Q}_{31}$	8.428256	1.270089
$\mathbf{Q}_{21}$	0.680756	0.366236	$\mathbf{Q}_{32}$	0.042292	0.027953
std err	0.002164	0.001370	std err	0.001335	0.000674
F_{fed}		11.33442	F_{fed}		18.02795
F_{mc}	30.46332	29.60259	F_{mc}	12.31672	19.53060

Table 2.1. Parameter Estimates

a, $\bar{\mathbf{r}}$, and σ, the standard error, and F statistics for various restrictions. The standard error is calculated in the usual way as the square root of the sum of squared errors (SSE) divided by the difference between the number of observations and the number of parameters, std err $= \text{SSE}/(n-k)$. The F statistic is also calculated in the usual way as

$$F = \frac{\text{SSE(restricted)} - \text{SSE(full)}}{\text{SSE(full)}} \times \frac{n-k}{r},$$

where r is the number of restricted parameters. Of course this statistic only has an F distribution with r and $n-k$ degrees of freedom in the linear case with linear restrictions and independent normally distributed residuals. However, the test is still useful for our situation since even with violations of linearity and normality the F distribution is approached asymptotically provided the parameter estimators are consistent, as they are for our model. In this case it is sometimes called a pseudo-F test. We consider a total of six scenarios: The Markov chain state space has 1, 2, or 3 dimensions and the short-term interest rate is proxied by the Federal Funds overnight rate or it is allowed to be freely estimated at each date by the regression.

The first thing to notice is that all of the F statistics in Table 2.1 are highly significant, having p-values of virtually zero in every case. This implies that all of the restrictions should be rejected, and the fullest model, which has a three dimensional state space and freely estimated initial short rate values at each date, is the best model even when the penalty for its unparsimoniousness is applied in the form of reduced degrees of freedom in the full model. A similar picture evolves when we look at each standard error. This statistic estimates the standard deviation of the residuals and also accounts for the degrees of freedom in the model. We see that the standard error is steadily reduced as more parameters enter the model.

We now turn our attention to the parameter estimates themselves. The rate of mean reversion, a, does behave as our intuition suggests it should. As we allow greater flexibility in the mean-reverting level, the rate at which this level is approached should increase. This is because with a fixed mean-reverting level, the rate of reversion will have to accommodate instances when the data diverges from the average. With a flexible mean-reverting level, these divergences can actually be considered instances of convergence to the more flexible level. From Table 2.1, we see that as we allow our Markov chain to have more states, the rate of mean reversion does increase.

On the other hand, estimates of the mean-reverting level, $\bar{r}$, are less economically intuitive. In the degenerate case, the level is quite reasonable at around 6.0 or 6.8%. However, when the Markov chain is allowed to switch between distinct states, the mean-reverting level tends to switch between unreasonably high and low values. When the short-term interest rate is restricted to be the Federal Funds rate, the mean-reverting level ranges between –24.4 and 27.5% for a two-state chain and –71.4 and 67.0% for a three-state chain. This is especially troubling when the high rate of mean reversion is also considered. We can see that restricting the initial short rate causes some of this problem, since when this constraint is relaxed, the mean-reverting levels become more reasonable. In particular, the two-state case switches between –1.7 and 10.5%, but the three-state case is still between –12.0 and 35.6%.

The volatility parameter, σ, turns out to be quite unimportant when the model is applied to the term structure data. For the degenerate case when the Markov chain has only one state, a 1% confidence interval is $0 \leq \sigma \leq 0.030851$. In fact, the sum of squared errors does not change perceptibly when the volatility is restricted to be $\sigma = 0$, and the p-value is virtually 1. A similar comment applies to the other cases, even in states one and two of the three state case, where the volatility is estimated to be somewhat larger, it is still not significantly different from zero. An explanation for this can be seen quite clearly in the degenerate case, which is Vasiček's [10] model. In this case the term structure equation can be written as

$$R_t(T) = \alpha(T)r_t + R_\infty\{1 - \alpha(T)\} + \frac{\sigma^2 T}{4a}\alpha(T)^2,$$

where $\alpha(T)$ is given previously and $R_\infty = \bar{r} - 0.5(\sigma/a)^2$, (see also Elliott and Kopp [6]). From this we can see that σ enters the formula through $(\sigma/a)^2$ and σ^2/a and with a small optimal volatility relative to the mean reversion rate, both of these quantities are small and likely to have little impact on the predicted yield to maturity. Although we do not have a closed-form solution for the more general cases, a similar reasoning may apply.

The entries of the transition rate matrix $\mathbf{Q}_{ij}$ can be interpreted as the rate at which the Markov chain is switching from state i to state j. It is perhaps easier to interpret these values if we convert them to monthly transition probabilities. We do this by calculating the exponential matrices $\mathbf{P} = e^{\mathbf{Q}t}$, with t taken to be one month (i.e. 1/12 of a year). The four cases with the number of states being two or three and the short-term interest rate being restricted or not are given as follows:

	$N = 2$	$N = 3$
Fed Fund	$\begin{bmatrix} 0.975384 & 0.024616 \\ 0.054456 & 0.945544 \end{bmatrix}$	$\begin{bmatrix} 0.717509 & 0.003359 & 0.279132 \\ 0.018137 & 0.961695 & 0.020168 \\ 0.411041 & 0.003423 & 0.585536 \end{bmatrix}$
Free Est.	$\begin{bmatrix} 0.982582 & 0.017418 \\ 0.029793 & 0.970207 \end{bmatrix}$	$\begin{bmatrix} 0.891514 & 0.002504 & 0.105982 \\ 0.018582 & 0.964250 & 0.017168 \\ 0.094595 & 0.002296 & 0.903109 \end{bmatrix}$

The i, j elements of these matrices represent the transition probabilities of going from state i to state j next month. From this, we can see that there is a fairly low probability of switching for either of the $N = 2$ cases, and a low probability of switching out of state 2 for the $N = 3$ cases, but there is a fairly high probability of switching from state 1 to 3 and from 3 to 1, especially for the restricted case when these probabilities are 27.9 and 41.1% respectively. Another informative quantity we can find is the steady state or limiting probabilities for each state. These limiting probabilities can be interpreted as the proportion of time spent in each state in the limit as time becomes large. It is easy to show that these transition matrices are associated with irreducible and ergodic Markov chains, so the limiting probabilities correspond to stationary probabilities, which satisfy the equations $\mathbf{P}^\mathsf{T}\pi = \pi$ and $\mathbf{1}^\mathsf{T}\pi = 1$. These steady state probabilities are given as follows (transposed as row vectors):

	$N = 2$	$N = 3$
Fed Fund	$\begin{bmatrix} 0.688688 & 0.311312 \end{bmatrix}$	$\begin{bmatrix} 0.546683 & 0.081187 & 0.372130 \end{bmatrix}$
Free Est.	$\begin{bmatrix} 0.631069 & 0.368931 \end{bmatrix}$	$\begin{bmatrix} 0.442306 & 0.062766 & 0.494928 \end{bmatrix}$

The last set of estimates for us to consider are the estimated Markov chain state probabilities and the estimated initial short-term interest rates. Rather than reporting these values in a tabular format, we present them graphically in Figure 2.1. We present the estimates for the three-state Markov chain. The top panel shows a graph of the probabilities for states 1, 2, and 3 through time, the middle panel shows a graph of the estimated short rate and the Federal Funds rate through time, and the lower panel shows a graph of the yields to maturity of the various zero-coupon bonds through time. The time scale of the graphs is matched to help draw inferences regarding how these three components are related to each other.

First we notice that the Markov chain state probabilities seem to behave as expected. Since according to Table 2.1, state 1 and state 3 are associated with a high and low mean-reverting level respectively, we expect the probability of being in state 1 to be higher and the probability of being in state 3 to be lower when rates are rising (and vice versa when rates are falling). This is generally observed when we compare the first and second panels. During 2002, we see the probability of state 2 rising and the probability of state 3 falling. Both of these states are associated with low mean-reverting levels, and therefore they should both correspond to falling interest rates as they do. However, state 3 has a much lower mean-reverting level than state 2, so state 2 should be associated with interest rates falling at a slower rate, which is consistent with what we observe between 2002 and 2004. As rates begin to rise again in 2004 and 2005, state 1 reasserts itself as the most likely state.

The second panel also allows us to compare the Federal Funds interest rate with the short rate filtered by the model. In general, they seem to agree quite well, although the filtered rate is usually slightly lower. Recall that this was also observed about the short maturity T-bills. There are a few substantial departures in 1999, 2000, and 2001, which likely account for most of the increase in sum of squared errors when we restrict the short-term interest rate. Indeed, when allowed to be freely estimated, the short rate is seen to follow the 1-month T-bill very closely.

The third panel graphs the data. We can see that this period is associated with a number of interesting phenomena. There are times when the interest rates are rising and the spread or slope of the yield curve is also increasing. The yield curve becomes very flat at the end of 2000 and beginning of 2001. The more common situation occurs when rates are falling but the spread is widening in 2001 and also the other common situation with rising rates and decreasing spread also occurs in 2004 and 2005. The most interesting feature

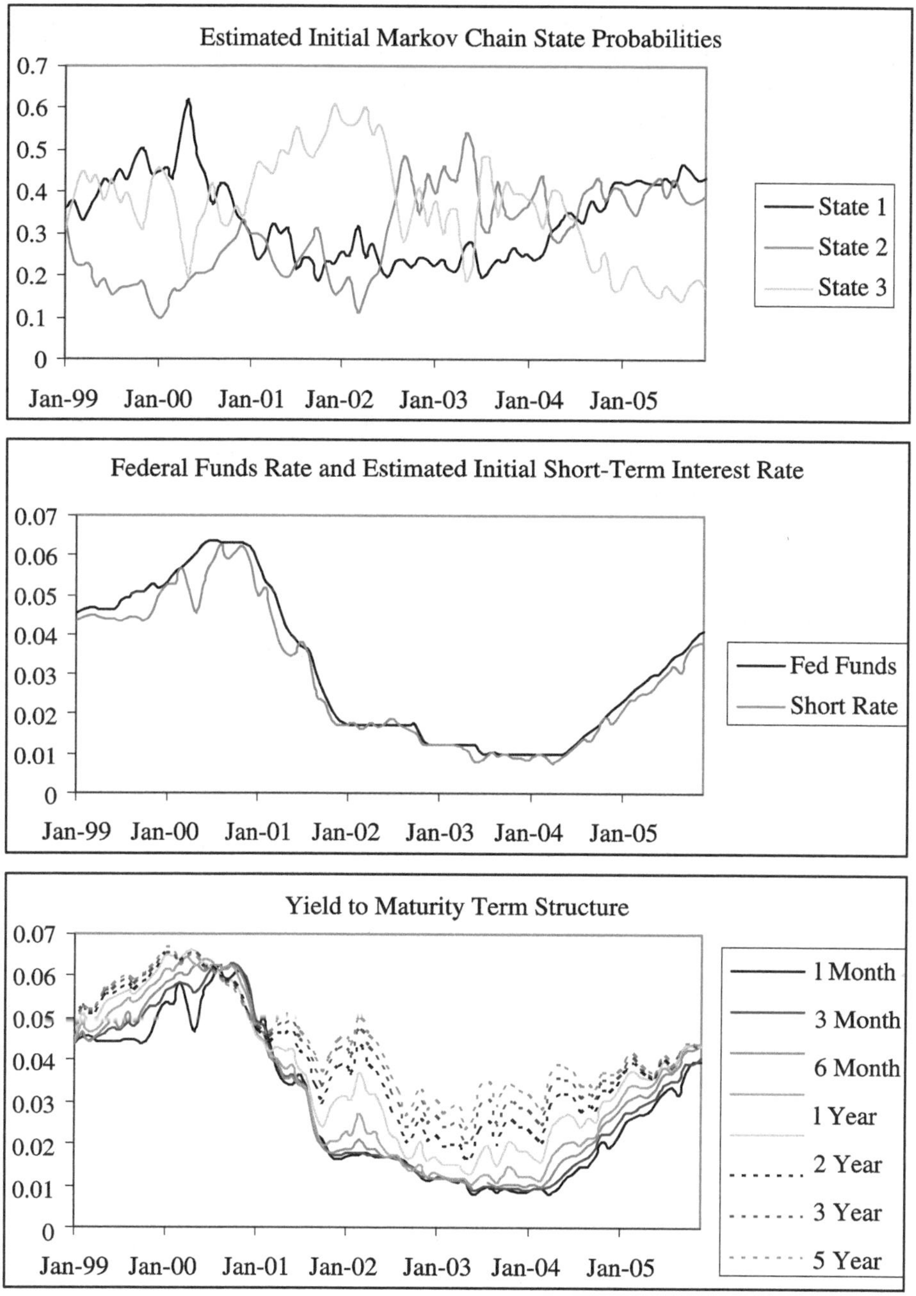

Fig. 2.1. State Probabilities and Initial Short-Term Interest Rate Estimates

occurs at the end of 2001 when the yield curve begins to twist with longer-term rates rising and short-term rates still falling. This is the time when a model with only one risk factor such as our degenerate case has the most difficulty in describing term structure dynamics, and this might explain why allowing the Markov chain as second factor of randomness to enter makes such a vast improvement in the models ability to explain this data.

2.5 Conclusion

We outline a methodology to incorporate a stochastic volatility and mean-reverting level into the short-term interest rate dynamics by using a Markov chain. We then show how to calculate the value of a zero-coupon bond. Using recent yield to maturity data, we estimate the model using a non-linear regression technique, and we find that the model makes a significant improvement in explaining the data over the basic model that excludes the Markov chain. Furthermore, we find that a three-state Markov chain makes a significant improvement over the two-state case. These improvements remain significant even when the initial short-rate is chosen optimally at each date, rather than being constrained to take values that proxy this rate such as the Federal Funds overnight rate. This suggests that models based on two-state regime switching may benefit from our more general N-state model construction.

References

1. Balduzzi, P., Das, S. R. and Foresi S. (1998). "The central tendency: A second factor in bond yields". *Review of Economics and Statistics*, 80(1): 62–72.
2. Cox, J. C., Ingersoll, J. E. and S. A. Ross (1985). "A theory of the term structure of interest rates". *Econometrica*, 53(2): 385–408.
3. Davidson, R. and J. G. MacKinnon (1993). *Estimation and Inference in Econometrics.* Oxford University Press.
4. Elliott, R. J. (1993). "New finite dimensional filters and smoothers for noisily observed Markov chains". *IEEE Transactions on Information Theory*, 39(1): 265–271.
5. Elliott, R. J. (1994). "Exact adaptive filters for Markov chains observed in Gaussian noise". *Automatica*, 30(9): 1399–1408.
6. Elliott, R. J. and P. E. Kopp. (1999). *Mathematics of Financial Markets.* Springer-Verlag.
7. Hull, J. and A. White (1990). "Pricing interest-rate derivative securities". *Review of Financial Studies*, 3(4): 573–592.
8. Naik, V. and M. H. Lee. (1997). "Yield curve dynamics with discrete shifts in economic regimes: Theory and estimation". Working Paper, University of British Columbia.
9. Roma, A. and W. Torous (1997). "The cyclical behavior of interest rates". *Journal of Finance*, 52(4): 1519–1542.
10. Vasiček, O. (1977). "An equilibrium characterization of the term structure". *Journal of Financial Economics*, 5(2): 177–188.

3

On Fair Valuation of Participating Life Insurance Policies With Regime Switching

Tak Kuen Siu

Department of Actuarial Mathematics and Statistics
School of Mathematical and Computer Sciences
Heriot-Watt University
Edinburgh, UK

Summary. We consider the valuation of participating life insurance policies using a regime-switching Esscher transform developed in Elliott, Chan and Siu (2005) when the market values of the reference asset are driven by a Markov-modulated geometric Brownian motion (GBM). We employ the Markov-modulated GBM driven by a continuous-time hidden Markov chain model to describe the impact of the switching behavior of the states of economy on the price dynamics of the reference asset. We also explore the change of measures technique to reduce the dimension of the valuation problem.

Key words: Participating policies; hidden Markov chain model; regime-switching Esscher transform.

3.1 Introduction

In recent years, participating life insurance products become more and more important in the finance and insurance markets due to their lower risk but provide comparable returns relative to other equity-index products. They are investment plans with associated life insurance benefits, a specified benchmark return, a guarantee of an annual minimum rate of return and a specified rule of the distribution of annual excess investment return above the guaranteed return. The policyholder has to pay a lump sum deposit to the insurer to initialize the contract. The insurer plays the role of a fund manager to manage and invest the funds in a specified reference portfolio. A major feature of these investment plans is the sharing of profits from an investment portfolio between the policyholders and the insurer. Typically, the insurer employs a specified rule of surplus distribution, namely, the reversionary bonus, to credit interest at or above a specified guaranteed rate to the policyholders every period, say per annum. If the surplus of the fund is positive at the maturity of the policy, the policyholders can also receive a terminal bonus. In the case that the insurer defaults at the maturity of the policy, the policyholders can only receive the outstanding assets. Grosen and Jørgensen [15] and Ballotta, Haberman and Wang [4] provided a comprehensive discussion on different contractual features of participating policies. Since there is a growing trend of using the market-based and fair valuation accountancy standards internationally for the implementation of risk management practice for participating policies, it is of practical importance and relevance to develop appropriate models for the valuation of these policies.

The pioneering work by Wilkie [22] introduced the use of the modern option pricing theory to investigate the embedded options in bonuses on with-profits life-insurance policies. Grosen and Jørgensen [15] developed a flexible contingent claims model to incorporate the minimum rate guarantee, bonus distribution and surrender risk. Priel, Putyatin and Nassar [20] incorporated the path dependence associated with the rule of the bonus distribution in their contingent claims model and adopted similarity transformations of variables to reduce the dimension of the problem. Bacinello ([1],[2]) adopted binomial schemes for computing the numerical solutions of the fair valuation problem of participating policies with various contractual features. Ballotta, Haberman and Wang [4] developed a valuation method for participating policies to incorporate reversionary bonus, terminal bonus and default option. Willder [23] adopted the modern option pricing approach for investigating the effects of various bonus strategies in unitized with-profit policies. Chu and Kwok [7] constructed a contingent claims model for participating policies that can incorporate rate guarrantee, bonuses and default risk.

In this paper, we consider the valuation of participating life insurance policies with bonus distributions and rate guarantees when the market values of the reference asset are driven by a Markov-modulated Geometric Brownian Motion (GBM). The switching behavior for the states of an economy can be attributed by the structural changes in economic conditions, political climates and business cycles, etc. Many life insurance products are relatively long dated and there can be substantial fluctuations in economic variables over a very long period of time. It is of practical importance and relevance to incorporate the switching behavior of the states of the economy for the valuation of participating policies. The market described by the Markov-modulated GBM model is incomplete, and, hence there are more than one equivalent martingale measures. The pioneering work by Gerber and Shiu [14] provided a pertinent solution to the option pricing problem in an incomplete market by using Esscher transform, a time-honored tool in actuarial science introduced by Esscher [10]. Here, we employ a modified version of the Esscher transform, namely, a regime-switching Esscher transform introduced in Elliott, Chan and Siu [9], to determine an equivalent martingale measure. This paper is outlined as follows.

Section 3.2 presents the fair valuation model for the partipating policies and a regime switching partial differential equation (PDE) for the valuation. Section

3.3 considers the problem of reducing the dimension of the regime-switching PDE using a change of probability measures. The final section suggests some potential topics for further investigation.

3.2 The model dynamics

We consider a financial model consisting of a risk-free money market account and a reference risky asset. We suppose that the market values of the reference asset are driven by a GBM with the drift and the volatility depending on the states of a continuous-time hidden Markov chain model. The states of the continuous-time hidden Markov chain model represents different states of an economy. We assume that the market is frictionless and that the mortality risk and surrender option are absent. We further impose certain assumptions on the rule of bonus distribution in our valuation model. In the sequel, we introduce the set-up of our model.

First, we fix a complete probability space $(\Omega, \mathcal{F}, \mathcal{P})$, where $\mathcal{P}$ is the real-world probability measure. Let $\mathcal{T}$ denote the time index set $[0, T]$. Let $\{W_t\}_{t \in \mathcal{T}}$ denote a standard Brownian motion on $(\Omega, \mathcal{F}, \mathcal{P})$ with respect to the $\mathcal{P}$-augmentation of its natural filtration $\mathcal{F}^W := \{\mathcal{F}_t^W\}_{t \in \mathcal{T}}$. The states of an economy augmented by a continuous-time Markov Chain process $\{X_t\}_{t \in \mathcal{T}}$ on $(\Omega, \mathcal{F}, \mathcal{P})$ with a finite state space $\mathcal{S} := (s_1, s_2 \dots, s_N)$. Without loss of generality, we can identify the state space of the process $\{X_t\}_{t \in \mathcal{T}}$ to be a finite set of unit vectors $\{e_1, e_2, \dots, e_N\}$, where $e_i = (0, \dots, 1, \dots, 0)^* \in \mathcal{R}^N$, where $*$ is a transpose of a vector. We suppose that the processes $\{X_t\}_{t \in \mathcal{T}}$ and $\{W_t\}_{t \in \mathcal{T}}$ are independent.

Write $Q(t)$ for the generator or Q-matrix $[q_{ij}(t)]_{i,j=1,2,\dots,N}$. Then, from Elliott, Aggoun and Moore [8], we have the following semi-martingale representation theorem for the process $\{X_t\}_{t \in \mathcal{T}}$:

$$X_t = X_0 + \int_0^t Q(s) X_s ds + M_t \ . \tag{3.1}$$

Here $\{M_t\}_{t \in \mathcal{T}}$ is an $\mathcal{R}^N$-valued martingale increment process with respect to the filtration generated by $\{X_t\}_{t \in \mathcal{T}}$.

Let $\{r(t, X_t)\}_{t \in \mathcal{T}}$ be the instantaneous market interest rate of a money market account, which depends on the state of the economy described by $\{X_t\}_{t \in \mathcal{T}}$; that is,

$$r(t, X_t) = \langle r, X_t \rangle \ , \ t \in \mathcal{T} \ , \tag{3.2}$$

where $r := (r_1, r_2, \dots, r_N)^*$ with $r_i > 0$ for each $i = 1, 2, \dots, N$ and $\langle \cdot, \cdot \rangle$ denotes the inner product in the space $\mathcal{R}^N$. For notational simplicity, we write r_t for $r(t, X_t)$.

In this case, the dynamics of the price process $\{B_t\}_{t\in\mathcal{T}}$ for the bank account is described by:

$$dB_t = r_t B_t dt \ , \quad B_0 = 1 \ . \tag{3.3}$$

Now, we assume that the expected growth rate $\{\mu_t\}_{t\in\mathcal{T}}$ and the volatility $\{\sigma_t\}_{t\in\mathcal{T}}$ of the market values of the asset also depend on $\{X_t\}_{t\in\mathcal{T}}$ and are described by:

$$\mu_t := \mu(t, X_t) = \langle \mu, X_t \rangle \ , \qquad \sigma_t := \sigma(t, X_t) = \langle \sigma, X_t \rangle \ , \tag{3.4}$$

where $\mu := (\mu_1, \mu_2, \ldots, \mu_N)$ and $\sigma := (\sigma_1, \sigma_2 \ldots, \sigma_N)$ with $\sigma_i > 0$ for each $i = 1, 2, \ldots, N$.

Then, the market values of the asset $\{A_t\}_{t\in\mathcal{T}}$ are governed by the Markov-modulated GBM with dynamics

$$dA_t = \mu_t A_t dt + \sigma_t A_t dW_t \ . \tag{3.5}$$

Let R_t denote the book value of the policy reserve and D_t the bonus reserve, at time $t \in \mathcal{T}$. Then, $A_t = R_t + D_t$, for each $t \in \mathcal{T}$. Note that $R(0) := \alpha A(0)$, $\alpha \in (0, 1]$, and $R(0)$ is interpreted as the single initial premium paid by the policyholder for acquiring the contract and α is the cost allocation parameter.

Write $c_R(A, R)$ for the interest rate credited to the policy reserve. Then, $c_R(A, R)$ is given by:

$$dR_t = c_R(A, R) R_t dt \ . \tag{3.6}$$

In practice, the specification of $c_R(A, R)$ depends on the rule of bonus distribution, which is decided by the management level of an insurance company. There is no consensus on a unified rule for the specification of $c_R(A, R)$. Typically, an insurer distributes to his/her policyholder a certain proportion, say δ, of the excess of the ratio of bonus reserve D_t to the policy reserve R_t over the target ratio β, which is a long-term constant specified by the management. The proportional constant δ is called the reversionary bonus distribution rate and it is assumed that $\delta \in (0, 1]$. For the crediting scheme of interest rate, it is also assumed that there is a specified guarantee rate r_g for the minimum interest rate credited to the policyholder's account. This means that $c_R(A, R) \geq r_g$. Grosen and Jørgensen [15], Prieul et al. [20] and Chu and Kwok [7] adopted different specifications for the interest rate crediting scheme. Here, we adopt the interest rate crediting scheme used in Chu and Kwok [7], which is

$$c_R(A_t, R_t) = \max\left(r_g, \ln\left(\frac{A_t}{R_t}\right) - \beta \right) \ . \tag{3.7}$$

It must be noted that the rate $c_R(A_t, R_t)$ is credited to the policy holder's account and depends on both β and r_g.

We adopt a martingale approach based on the Esscher transform to determine the fair value of a participating policy. Bacinello [1] pioneered the use of the martingale approach for the valuation of participating policies. The fair price of the participating policy is expressed as the expectation of the discounted payoff of the policy under the risk-neutral equivalent martingale measure $\mathcal{Q}$, under which the process of the discounted prices of any security is a martingale (see Harrison and Kreps [16]). Ballotta [3] employed the constant-parameter Esscher transform to determine an equivalent martingale pricing measure in an incomplete market described by the jump-diffusion process. In the sequel, we describe the use of the regime-switching Esscher transform for determining an equivalent martingale measure.

Let Y_t denote the logarithmic return $\ln(A_t/A_0)$ from the asset over the time duration $[0,t]$. Write $\{\mathcal{F}_t^X\}_{t\in\mathcal{T}}$ and $\{\mathcal{F}_t^Y\}_{t\in\mathcal{T}}$ for the $\mathcal{P}$-augmentation of the natural filtrations generated by $\{X_t\}_{t\in\mathcal{T}}$ and $\{Y_t\}_{t\in\mathcal{T}}$, respectively. For each $t\in\mathcal{T}$, we define $\mathcal{G}_t$ as the σ-algebra $\mathcal{F}_t^X \vee \mathcal{F}_t^Y$. Let $\theta(t,X_t)$ be the regime-switching Esscher parameter, which depends on X_t. $\theta(t,X_t)$ can be written as follows:

$$\theta(t,X_t) = \langle \theta, X_t\rangle = \sum_{i=1}^{N} \theta_i \langle X_t, e_i\rangle \ , \tag{3.8}$$

where $\theta := (\theta_1,\theta_2,\dots,\theta_N)^* \in \mathcal{R}^N$. We write θ_t for $\theta(t,X_t)$.

Then, as in Elliott, Chan and Siu [9], the regime-switching Esscher transform $\mathcal{Q}_\theta$ equivalent to $\mathcal{P}$ on $\mathcal{G}_t$ is defined by

$$\left.\frac{d\mathcal{Q}_\theta}{d\mathcal{P}}\right|_{\mathcal{G}_t} = \frac{\exp\left(\int_0^t \theta_s dY_s\right)}{E_{\mathcal{P}}\left[\exp\left(\int_0^t \theta_s dY_s\right)\,\middle|\,\mathcal{F}_t^X\right]} \ , \qquad t\in\mathcal{T} \ . \tag{3.9}$$

The Radon-Nikodym derivative of the regime-switching Esscher transform is given by (see Elliott, Chan and Siu, [9]):

$$\left.\frac{d\mathcal{Q}_\theta}{d\mathcal{P}}\right|_{\mathcal{G}_t} = \exp\left(\int_0^t \theta_s\sigma_s dW_s - \frac{1}{2}\int_0^t \theta_s^2\sigma_s^2 ds\right) \ . \tag{3.10}$$

Write $\{\tilde{\theta}_t\}_{t\in\mathcal{T}}$ for a family of risk-neutral regime-switching Esscher parameters. By the fundamental theorem of asset pricing (see Harrison and Kreps [16], Harrison and Pliska ([17],[18])), the absence of arbitrage opportunities is "essentially" equivalent to the existence of an equivalent martingale measure under which the discounted stock price process is a martingale. Here, the martingale condition is given by considering an enlarged filtration as follows:

$$A_0 = E_{\mathcal{Q}_{\tilde{\theta}}}\left[\exp\left(-\int_0^t r_s ds\right)A_t\,\middle|\,\mathcal{F}_t^X\right] \ , \quad \text{for any } t\in\mathcal{T} \ . \tag{3.11}$$

As in Elliott, Chan and Siu [9], we can determine $\tilde{\theta}_t$ uniquely from the martingale condition:

$$\tilde{\theta}_t = \frac{r_t - \mu_t}{\sigma_t^2} = -\frac{\lambda_t}{\sigma_t} = \sum_{i=1}^{N} \left(-\frac{\lambda_i}{\sigma_i} \right) \langle X_t, e_i \rangle \ , \qquad t \in \mathcal{T} \ , \tag{3.12}$$

where $\lambda_t := \frac{\mu_t - r_t}{\sigma_t}$ is the market price of risk or unit risk premium of the reference asset at time t; $\lambda_i := \frac{\mu_i - r_i}{\sigma_i}$, for each $i = 1, 2, \ldots, N$.

Let $\tilde{\mathcal{G}}_t$ denote the enlarged σ-field $\mathcal{F}_T^X \vee \mathcal{F}_t^Y$, for any $t \in \mathcal{T}$. From the martingale condition, the Radon-Nikodym derivative of the risk-neutral regime switching Esscher measure $\mathcal{Q}_{\tilde{\theta}}$ is given by:

$$\left. \frac{d\mathcal{Q}_{\tilde{\theta}}}{d\mathcal{P}} \right|_{\mathcal{G}_t} = \exp \left[\int_0^t \left(\frac{r_s - \mu_s}{\sigma_s} \right) dW_s - \frac{1}{2} \int_0^t \left(\frac{r_s - \mu_s}{\sigma_s} \right)^2 ds \right] . \tag{3.13}$$

By Girsanov's theorem, the process $\tilde{W}_t = W_t + \int_0^t (\frac{r_s - \mu_s}{\sigma_s}) ds$ is a standard Brownian motion with respect to $\{\tilde{\mathcal{G}}_t\}_{t \in \mathcal{T}}$ under $\mathcal{Q}_{\tilde{\theta}}$. Hence, the market values of the asset under $\mathcal{Q}_{\tilde{\theta}}$ can be written as

$$dA_t = r_t A_t dt + \sigma_t A_t d\tilde{W}_t \ . \tag{3.14}$$

Suppose $V(A, R, X, t)$ denotes the value of the participating policy at time t. The terminal payoff of the participating policy $V(A, R, X, T)$ on the policy maturity date T, when $X_T = X$, is given by:

$$V(A, R, X, T) = \begin{cases} A_T & \text{if } A_T < R_T \\ R_T & \text{if } R_T \leq A_T \leq \frac{R_T}{\alpha} \\ R_T + \gamma P_{1T} & \text{if } A_T > \frac{R_T}{\alpha} \end{cases} \tag{3.15}$$

where γ is the terminal bonus distribution rate and $P_{1T} := \max(\alpha A_T - R_T, 0)$ is the terminal bonus option.

Let $P_{2T} := \max(R_T - A_T, 0)$, where P_{2T} represents the terminal default option on the policy maturity date T. Then, it can be shown that the terminal payoff $V(A, R, X, T)$ can be written in the following form:

$$V(A, R, X, T) = R_T + \gamma P_{1T} - P_{2T} \ . \tag{3.16}$$

Suppose the trajectory of the hidden process X from time 0 to time T is given in advance. Then, the fair value of the participating policy $V(A, R, t \mid \tilde{\mathcal{G}}_t)$ at time t is given by

$$V(A, R, t \mid \tilde{\mathcal{G}}_t) = E_{\mathcal{Q}_{\tilde{\theta}}} \left[\exp\left(- \int_t^T r_s ds \right) V(A, R, X, T) \,\Big|\, \tilde{\mathcal{G}}_t \right] . \tag{3.17}$$

The pricing result can be justified by minimizing the relative entropy of an equivalent martingale measure and the real-world probability $\mathcal{P}$ (see Miyahara [19] and Elliott, Chan and Siu [9]).

Note that

$$R_T = R_t \exp\left(\int_t^T c_R(A_s, R_s)ds\right)$$
$$= R_t \exp\left[\int_t^T \max\left(r_g, \ln\left(\frac{A_s}{R_s}\right) - \beta\right)ds\right] . \quad (3.18)$$

Since one cannot determine which of $\ln(\frac{A_s}{R_s}) - \beta$ or r_g is greater when the regime X_s is fixed, it is difficult, if not impossible, to write the fair value of the participating policy in closed form as some sort of expectation using the joint probability density function or characteristic function of the occupation times of the Markov chain X model. In this case, the PDE approach can provide a convenient way to evaluate the fair value of the participating policy.

For determining the value of the participating policy, we consider an additional state variable R_t, which is a path integral of the process A. Now, if $A_t = A$, $R_t = R$ and $X_t = X$ are given at time t, the value of the participating policy $V(A, R, X, t)$ at time t is given by:

$$V(A, R, X, t)$$
$$= E_{\mathcal{Q}_{\tilde{\theta}}}\left[\exp\left(-\int_t^T r_s ds\right)V(A, R, X, T) \,\middle|\, A_t = A, R_t = R, X_t = X\right] . \quad (3.19)$$

Write $\tilde{V}(A, R, X, t)$ for $\exp(-\int_0^t r_s ds)V(A, R, X, t)$. Since R_t is a path integral of A_t and A_t is a Markov process given that the trajectory of X is known, (A_t, R_t) is a two-dimensional Markov process given the known trajectory of X. Due to the fact that X is also a Markov process, (A_t, R_t, X_t) is a three-dimensional Markov process with respect to the enlarged information set $\mathcal{G}_t$, where $\mathcal{G}_t = \sigma\{A_u, X_u | u \in [0, t]\}$. Then, by the Markov property of (A_t, R_t, X_t),

$$\tilde{V}(A, R, X, t) = E_{\mathcal{Q}_{\tilde{\theta}}}\left[\exp\left(-\int_0^T r_s ds\right)V(A, R, X, T) \,\middle|\, \mathcal{G}_t\right] . \quad (3.20)$$

Hence, $\tilde{V}(A, R, X, t)$ is a $\mathcal{G}_t$-martingale under $\mathcal{Q}_{\tilde{\theta}}$.

Let $\tilde{\mathbf{V}}(A, R, t)$ denote the N-dimensional vector

$$(\tilde{V}(A, R, e_1, t), \ldots, \tilde{V}(A, R, e_N, t)) .$$

Then, $\tilde{V}(A, R, X, t) = \left\langle \tilde{\mathbf{V}}(A, R, t), X_t \right\rangle$.

Write $c_R(u)$ for $c_R(A_u, R_u)$. Then, by applying Itô's differentiation rule to $\tilde{V}(A, R, X, t)$,

$$\tilde{V}(A,R,X,t) = \tilde{V}(A,R,X,0) + \int_0^t \left(\frac{\partial \tilde{V}}{\partial u} + r_u A_u \frac{\partial \tilde{V}}{\partial A} + \frac{1}{2}\sigma_u^2 A_u^2 \frac{\partial^2 \tilde{V}}{\partial A^2} \right.$$
$$\left. + c_R(u) R_u \frac{\partial \tilde{V}}{\partial R} \right) du + \int_0^t \frac{\partial \tilde{V}}{\partial A} \sigma_u A_u dW_u + \int_0^t \left\langle \tilde{\mathbf{V}}, dX_u \right\rangle , \tag{3.21}$$

and $dX_t = Q(t)X_t dt + dM_t$.

Since $\tilde{V}(A,R,X,t)$ is a $\mathcal{G}_t$-martingale under $\mathcal{Q}_{\tilde{\theta}}$, all terms with bounded variation must be identical to zero. Hence, we obtain the PDE for the discounted fair value $\tilde{V}(A,R,X,t)$:

$$\frac{\partial \tilde{V}}{\partial t} + r_t A_t \frac{\partial \tilde{V}}{\partial A} + \frac{1}{2}\sigma_t^2 A_t^2 \frac{\partial^2 \tilde{V}}{\partial A^2} + c_R(t) R_t \frac{\partial \tilde{V}}{\partial R} + \left\langle \tilde{\mathbf{V}}, QX \right\rangle = 0 . \tag{3.22}$$

Let $\mathbf{V}(A,R,t)$ denote the N-dimensional vector

$$\big(V(A,R,e_1,t), \ldots, V(A,R,e_N,t)\big).$$

Define the partial differential operator $\mathcal{L}_{A,R,X}$ as follows:

$$\mathcal{L}_{A,R,X}(V,t) = -r_t V + \frac{\partial V}{\partial t} + r_t A_t \frac{\partial V}{\partial A} + \frac{1}{2}\sigma_t^2 A_t^2 \frac{\partial^2 V}{\partial A^2} + c_R(t) R_t \frac{\partial V}{\partial R} . \tag{3.23}$$

Then, as in Buffington and Elliott ([5],[6]), $V(A,R,X,t)$ satisfies the PDE

$$\mathcal{L}_{A,R,X}(V,t) + \langle \mathbf{V}, QX \rangle = 0 , \tag{3.24}$$

with the terminal condition

$$V(A,R,X,T) = R_T + \gamma \max(\alpha A_T - R_T, 0) - \max(R_T - A_T, 0) . \tag{3.25}$$

3.3 Dimension reduction to regime-switching PDE

Chu and Kwok [7] adopted the method of similarity transformations to reduce the dimension of the PDE for the valuation of a participating policy without switching regimes. Here, we use a change of probability measures to reduce the dimension of the regime-switching PDE in the last section. The regime-switching PDE derived from the change of measures depends on two state variables including a new observable state variable and the state of the economy. When there is no regime switching, the PDE derived from the change of measures resembles to the one obtained from the method of similarity transformations by Chu and Kwok [7]. By employing the method in Buffington and Elliott ([5],[6]), we further simplify the problem by writing the regime-switching PDE as a system of coupled PDEs without switching regimes.

First, we define a new observable state variable $Z := \ln\left(\frac{A}{R}\right)$ and a function $U(Z, X, t) := V(A, R, X, t)/R$. We assume that

$$c_Z(Z) = c_R(A, R), \quad V_Z(Z, X, T) = \frac{V(A, R, X, T)}{R_T} . \tag{3.26}$$

Note that the terminal condition can be written as

$$V_Z(Z, X, T) = 1 + \gamma \max(\alpha e^{Z_T} - 1, 0) - \max(1 - e^{Z_T}, 0) . \tag{3.27}$$

By Itô's lemma, the dynamics of the new state variable Z under $\mathcal{P}$ are given by

$$dZ_t = \left(\mu_t - c_Z(Z_t) - \frac{1}{2}\sigma_t^2\right) dt + \sigma_t dW_t . \tag{3.28}$$

Now, we define a $\mathcal{Q}_{\tilde{\theta}}$-martingale with respect to $\mathcal{G}_t$ as

$$\xi(t) = \exp\left(-\int_0^t r_s ds\right) \frac{A_t}{A_0} . \tag{3.29}$$

Under $\mathcal{Q}_{\tilde{\theta}}$, the dynamics of A_t are given by

$$A_t = A_0 \exp\left[\int_0^t \left(r_s - \frac{1}{2}\sigma_s^2\right) ds + \int_0^t \sigma_s dW_s\right] . \tag{3.30}$$

Hence,

$$\xi(t) = \exp\left(-\int_0^t \frac{1}{2}\sigma_s^2 ds + \int_0^t \sigma_s dW_s\right) . \tag{3.31}$$

Then, we define a new equivalent measure $\hat{Q}$ as

$$\left.\frac{d\hat{Q}}{d\mathcal{Q}_{\tilde{\theta}}}\right|_{\mathcal{G}_t} = \xi(t) , \quad t \in \mathcal{T} . \tag{3.32}$$

By Girsanov's theorem, the process

$$\hat{W}_t := \tilde{W}_t - \int_0^t \sigma_s ds , \tag{3.33}$$

is a standard Brownian motion under $\hat{Q}$ with respect to $\mathcal{G}_t$.

Under $\hat{Q}$, the dynamics of A can be represented by

$$dA_t = (r_t + \sigma_t^2) A_t dt + \sigma_t A_t d\hat{W}_t . \tag{3.34}$$

Therefore, under $\hat{Q}$, the dynamics of Z_t are given by

$$dZ_t = \left(r_t + \frac{1}{2}\sigma_t^2 - c_Z(Z_t)\right)dt + \sigma_t d\hat{W}_t \ . \tag{3.35}$$

By Bayes' rule,

$$\begin{aligned}\tilde{V}(A,R,X,t) &= E_{\mathcal{Q}_{\tilde{\theta}}}\left[\exp\left(-\int_0^T r_s ds\right)V(A,R,X,T) \,\Big|\, \mathcal{G}_t\right] \\ &= E_{\hat{Q}}\left[\frac{\xi(t)}{\xi(T)}\exp\left(-\int_0^T r_s ds\right)V(A,R,X,T) \,\Big|\, \mathcal{G}_t\right] \\ &= \exp\left(-\int_0^t r_s ds\right)A_t E_{\hat{Q}}\left[\left(\frac{R_T}{A_T}\right)\frac{V(A,R,X,T)}{R_T} \,\Big|\, \mathcal{G}_t\right] \\ &= \exp\left(-\int_0^t r_s ds\right)A_t E_{\hat{Q}}(e^{-Z_T}V_Z(Z,X,T) \mid \mathcal{G}_t) \ . \end{aligned} \tag{3.36}$$

Write $\tilde{V}_Z(Z,X,t)$ for $E_{\hat{Q}}(e^{-Z_T}V_Z(Z,X,T)|\mathcal{G}_t)$. Then, $\tilde{V}_Z(Z,X,t)$ is a $\hat{Q}$-martingale with respect to $\mathcal{G}_t$. It is worth mentioning that all asset prices are $\hat{Q}$-martingale with respect to $\mathcal{G}_t$ when discounted by the asset market value A. In this case, all asset prices are measured in units of A. We call the market value A of the asset a numeráire. Consequently, changing the measure from $\mathcal{Q}_{\tilde{\theta}}$ to $\hat{Q}$ is equivalent to the change of numeráire from the price process of the money market account B to the price process of the reference asset A.

Let $\tilde{\mathbf{V}}_Z(Z,t)$ denote the N-dimensional vector $(\tilde{V}_Z(Z,e_1,t),\ldots,\tilde{V}_Z(Z,e_N,t))$. Then, $\tilde{V}_Z(Z,X,t) = \left\langle \tilde{\mathbf{V}}_Z(Z,t), X_t\right\rangle$. Again, by applying Itô's differentiation rule to $\tilde{V}_Z(Z,X,t)$ we obtain

$$\tilde{V}_Z(Z,X,t) = \tilde{V}_Z(Z,X,0) \tag{3.37}$$

$$+\int_0^t \left[\frac{\partial \tilde{V}_Z}{\partial u} + \left(r_u + \frac{1}{2}\sigma_u^2 - c_Z(Z_u)\right)\frac{\partial \tilde{V}_Z}{\partial Z} + \frac{1}{2}\sigma_u^2\frac{\partial^2 \tilde{V}_Z}{\partial Z^2}\right]du$$

$$+\int_0^t \frac{\partial \tilde{V}_Z}{\partial Z}\sigma_u d\hat{W}_u + \int_0^t \left\langle \tilde{\mathbf{V}}_{\mathbf{Z}}, dX_u\right\rangle \ , \tag{3.38}$$

and $dX_t = Q(t)X_t dt + dM_t$.

Note that $\tilde{V}_Z(Z,X,t)$ is a $\mathcal{G}_t$-martingale under $\hat{Q}$ and all terms with bounded variation must be identical to zero. Hence, we get the following PDE with one observable state variable Z for $\tilde{V}_Z(Z,X,t)$:

$$\frac{\partial \tilde{V}_Z}{\partial t} + \left(r_t + \frac{1}{2}\sigma_t^2 - c_Z(Z_t)\right)\frac{\partial \tilde{V}_Z}{\partial Z} + \frac{1}{2}\sigma_t^2\frac{\partial^2 \tilde{V}_Z}{\partial Z^2} + \left\langle \tilde{\mathbf{V}}_{\mathbf{Z}}, X_t\right\rangle = 0 \ . \tag{3.39}$$

Notice that

$$U(Z,X,t) = e^{Z_t}\tilde{V}_Z(Z,X,t) \ , \tag{3.40}$$

so that

$$\frac{\partial \tilde{V}_Z}{\partial t} = e^{-Z_t} \frac{\partial U}{\partial t} , \quad \frac{\partial \tilde{V}_Z}{\partial Z} = e^{-Z_t} \left(\frac{\partial U}{\partial Z} - U \right) ,$$
$$\frac{\partial^2 \tilde{V}_Z}{\partial Z^2} = e^{-Z_t} \left(\frac{\partial^2 U}{\partial Z^2} - 2\frac{\partial U}{\partial Z} + U \right) . \tag{3.41}$$

Define the partial differential operator $\mathcal{L}_{Z,X}$ as follows:

$$\mathcal{L}_{Z,X}(U,t) = \frac{\partial U}{\partial t} + \left(r_t - \frac{1}{2}\sigma_t^2 - c_Z(Z_t) \right) \frac{\partial U}{\partial Z} + \frac{1}{2}\sigma_t^2 \frac{\partial^2 U}{\partial Z^2} - (r_t - c_Z(Z_t))U . \tag{3.42}$$

Then, the process U satisfies the PDE

$$\mathcal{L}_{Z,X}(U,t) + \langle U, X_t \rangle = 0 , \tag{3.43}$$

with the auxillary condition

$$U(Z,X,T) = 1 + \gamma \max(\alpha e^{Z_T} - 1, 0) - \max(1 - e^{Z_T}, 0) . \tag{3.44}$$

Following Buffington and Elliott ([5],[6]), we reduce the above regime-switching PDE to a system of N coupled PDEs without regime switching with $X_t = e_1, e_2, \ldots, e_N$. First, we suppose that $X_t = e_i$ $(i = 1, 2, \ldots, N)$. Then,

$$r_t = \langle r, X_t \rangle = \langle r, e_i \rangle = r_i ,$$
$$\sigma_t = \langle \sigma, X_t \rangle = \langle \sigma, e_i \rangle = \sigma_i . \tag{3.45}$$

Let $U_i := U(Z, e_i, t)$, for $i = 1, 2 \ldots, N$. Write $\mathbf{U}$ for $(U_1, U_2, \ldots, U_N)$. Define the following partial differential operator $\mathcal{L}_{Z,e_i}$, for $i = 1, 2, \ldots, N$:

$$\mathcal{L}_{Z,e_i}(U_i,t) = \frac{\partial U_i}{\partial t} + \left(r_i - \frac{1}{2}\sigma_i^2 - c_Z(Z_t) \right) \frac{\partial U_i}{\partial Z} + \frac{1}{2}\sigma_i^2 \frac{\partial^2 U_i}{\partial Z^2} - (r_i - c_Z(Z_t))U_i . \tag{3.46}$$

Then, the N-dimensional vector $\mathbf{U}$ satisfies the following system of N coupled PDEs without regime switching:

$$\mathcal{L}_{Z,e_i}(U_i,t) + \langle \mathbf{U}, A e_i \rangle = 0 , \tag{3.47}$$

with the auxillary condition

$$U(Z,e_i,T) = 1 + \gamma \max(\alpha e^{Z_T} - 1, 0) - \max(1 - e^{Z_T}, 0) . \tag{3.48}$$

3.4 Further investigation

For further investigation, one may consider the valuation of with-profits insurance products when the market values of the reference portfolio are governed by other types of regime-switching models, such as the Markov-modulated Lévy processes. It is interesting to explore the use of other techniques for choosing an equivalent martingale measure in the literature, such as minimizing the quadratic utility by Föllmer and Sondermann [11], Föllmer and Schweizer [12] and Schweizer [21] and the quantile-based hedging by Föllmer and Leukert [13], etc., for the valuation of with-profits insurance products under the Markov-modulated diffusion processes and other specifications of the market values of the reference portfolio.

References

1. Bacinello, A. R. (2001). "Fair pricing of life insurance participating policies with a minimum interest rate guaranteed". *ASTIN Bulletin*, 31 (2): 275-297.
2. Bacinello, A. R. (2003). "Fair valuation of a guaranteed life insurance participating contract embedding a surrender option". *Journal of Risk and Insurance*, 70(3): 461-487.
3. Ballotta, L. (2005). "A Lévy process-based framework for the fair valuation of participating life insurance contracts". *Insurance: Mathematics and Economics*, 37(2): 173-96.
4. Ballotta, L., Haberman, S. and N. Wang (2006). "Guarantees in with-profit and unitised with profit life insurance contracts: fair valuation problem in presence of the default option". *Journal of Risk and Insurance*, 73 (1): 97-121.
5. Buffington, J. and R. J. Elliott (2001). "Regime switching and European options". *In Stochastic Theory and Control, Proceedings of a Workshop, Lawrence, K.S.*, October, Springer Verlag, 73-81.
6. Buffington, J. and R. J. Elliott (2002). "American options with regime switching". *International Journal of Theoretical and Applied Finance*, 5: 497-514.
7. Chu, C. C. and Y. K. Kwok (2006). "Pricing participating policies with rate guarantees and bonuses". *International Journal of Theoretical and Applied Finance*, To appear.
8. Elliott, R. J., Aggoun, L. and J. B. Moore (1994). *Hidden Markov Models: Estimation and Control.* Springer-Verlag.
9. Elliott, R. J., Chan, L. L., and T. K. Siu (2005). "Option pricing and Esscher transform under regime switching". *Annals of Finance*, 1(4): 423-432.
10. Esscher, F. (1932). "On the probability function in the collective theory of risk". *Skandinavisk Aktuarietidskrift*, 15: 175-195.

11. Föllmer, H. and D. Sondermann (1986). "Hedging of contingent claims under incomplete information. In: Hildenbrand, W. and Mas-Colell, A. (eds.)", *Contributions to Mathematical Economics.* North Holland. 205-223.
12. Föllmer, H. and M. Schweizer (1991) "Hedging of contingent claims under incomplete information. In: Davis, M.H.A., and Elliott, R.J. (eds.)", *Applied Stochastic Analysis.* Gordon and Breach. 389-414.
13. Föllmer, H. and P. Leukert (1999). "Quantile hedging". *Finance and Stochastics*, 3(3): 251-273.
14. Gerber, H. U. and E. S. W. Shiu (1994). "Option pricing by Esscher transforms (with discussions)". *Transactions of the Society of Actuaries*, 46: 99-191.
15. Grosen, A. and P. L. Jørgensen (2000). "Fair valuation of life insurance liabilities: The impact of interest rate guarantees, surrender options, and bonus policies". *Insurance: Mathematics and Economics*, 26: 37-57.
16. Harrison, J. M. and D. M. Kreps (1979). "Martingales and arbitrage in multiperiod securities markets". *Journal of Economic Theory*, 20: 381-408.
17. Harrison, J. M. and S. R. Pliska (1981). "Martingales and stochastic integrals in the theory of continuous trading". *Stochastic Processes and Their Applications*, 11: 215-280.
18. Harrison, J. M. and S. R. Pliska (1983). "A stochastic calculus model of continuous trading: Complete markets". *Stochastic Processes and Their Applications*, 15: 313-316.
19. Miyahara, Y. (2001). "Geometric Lévy process and MEMM: Pricing model and related estimation problems". *Asia-Pacific Financial Markets*, 8: 45-60.
20. Prieul, D., Putyatin, V. and T. Nassar (2001). "On pricing and reserving with-profits life insurance contracts". *Applied Mathematical Finance*, 8: 145-166.
21. Schweizer, M. (1996). "Approximation pricing and the variance-optimal martingale measure". *Annals of Probability*, 24: 206-236.
22. Wilkie, A. D. (1987). "An option pricing approach to bonus policy". *Journal of the Institute of Actuaries*, 114: 21-77.
23. Willder, M. (2004). "An option pricing approach to pricing guarantees given under unitised with-profits policies". *Ph.D. Thesis*, Heriot-Watt University.

4

Pricing Options and Variance Swaps in Markov-Modulated Brownian Markets

Robert J. Elliott[1] and Anatoliy V. Swishchuk[2]

[1] Haskayne School of Business
University of Calgary
2500 University Drive NW
Calgary, Alberta, Canada T2N 1N4
relliott@ucalgary.ca

[2] Department of Mathematics and Statistics
University of Calgary
2500 University Drive NW
Calgary, Alberta, Canada T2N 1N4
aswish@math.ucalgary.ca

Summary. A Markov-modulated market consists of a riskless asset or bond, B, and a risky asset or stock, S, whose dynamics depend on Markov process x. We study the pricing of options and variance swaps in such markets. Using the martingale characterization of Markov processes, we note the incompleteness of Markov-modulated markets and find the minimal martingale measure. Black-Scholes formulae for Markov-modulated markets with or without jumps are derived. Perfect hedging in a Markov-modulated Brownian and a fractional Brownian market is not possible as the market is incomplete. Following the idea proposed by Föllmer and Sondermann [13] and Föllmer and Schweizer [12]) we look for the strategy which locally minimizes the risk. The residual risk processes are determined in these situations. Variance swaps for stochastic volatility driven by Markov process are also studied.

Key words: Markov-modulated markets with jumps, option pricing, variance swaps, minimal martingale measure

4.1 Introduction

Consider a standard probability space (Ω, F, F_t, P) with a right-continuous, complete filtration F_t and probability P.

A *Brownian* (B, S)*-security market* will consist of a riskless asset, (bond or bank account) $B = \{B_t, t \geq 0\}$, and risky asset, (stock or share), $S = \{S_t, t \geq$

$0\}$ whose dynamics are given by the two equations:

$$\begin{cases} dB_t = rB_t, & B_0 > 0, r > 0, \\ dS_t = S_t(\mu dt + \sigma dw_t), & S_0 > 0,\ \sigma > 0,\ \mu \in R. \end{cases} \tag{4.1}$$

Here r denotes the instanteneous interest rate, μ the appreciation rate, σ the volatility and $w = \{w_t, t \geq 0\}$ is an F_t-Brownian motion on (Ω, F, F_t, P). $Ew_t = 0, \quad Ew_t^2 = t$, where E is an expectation with respect to the measure P. It is well-known that this market has no arbitrage and is complete (see Elliott and Kopp [9]).

However, even in the Brownian motion framework, there is an arbitrage opportunity if Stratonovich integration is used in the definition of self-financing portfolios. Consider an example, due to Rogers [26] and Shiryaev [27]. Let $B_t = e^{rt}$ and $S_t = e^{rt+w_t}$ represent the bond price and stock price, respectively, at time t. Then (B_t, S_t), with $\cdot dw_t$ denoting the Stratonovich differential, constitutes a Black-Scholes market, namely:

$$\begin{cases} dB_t = rB_t dt \\ dS_t = S_t(rdt + \cdot dw_t). \end{cases} \tag{4.2}$$

Consider the portfolio $\pi_t = (\beta_t, \gamma_t)$, where

$$\begin{cases} \beta_t = 1 - e^{2w_t}, \\ \gamma_t = 2(e^{w_t} - 1). \end{cases} \tag{4.3}$$

Then using (4.2) the value at time t of this portfolio is

$$X_t^\pi = \beta_t B_t + \gamma_t S_t = e^{rt}[e^{w_t} - 1]^2.$$

From (4.2) and (4.3) it is easy to check that

$$dX_t^\pi = \beta_t dB_t + \gamma_t \cdot dS_t.$$

Thus, π_t is self-financing if one replaces the Itô integral by the Stratonovich integral in the definition of self-financing. Also, $X_0^\pi = 0, \quad X_t^\pi \geq 0$ for $t > 0$, and $EX_t^\pi > 0$ for $t > 0$, and so there is arbitrage in this model.

Remark 1. It is well-known, that if Itô integrals are used in (4.1) there is no arbitrage. Such markets are also complete (see Elliott & Kopp [9]).

Remark 2. Note that if the Stratonovich integral is used in the definition of self-financing portfolios of a Brownian market with jumps then there is arbitrage.

A *Markov-modulated Brownian (B, S)-security market* consists of a riskless asset, (a bond or bank account), B, and risky asset, (a stock or share), S, which satisfy the following system of two equations:

$$\begin{cases} dB_t = r(x_t)B_t, \quad B_0 > 0, r(x) > 0, \\ dS_t = S_t(\mu(x_t)dt + \sigma(x_t)dw_t), \quad S_0 > 0,\, \sigma(x) > 0,\, \mu(x) \in R. \end{cases} \tag{4.4}$$

Here $r(x)$ is an interest rate, $\mu(x)$ is an appreciation rate, $\sigma(x)$ is a volatility. They are bounded continuous functions on a locally compact metric space X, the state space of a continuous-time Markov process $x = \{x_t, t \geq 0\}$, $x_0 = x$. As before, w is a Brownian motion independent of x, and the second stochastic differential equation is an Itô equation.

The main goal of this paper is to study model (4.4), including such models with jumps.

The paper is organized as follows. Literature review is provided in Section 4.2. Using the martingale characterization of Markov processes (Section 4.3), we state the incompleteness of Markov-modulated Brownian (B, S)-security markets (4.4) without, (Section 4.4), and with, (Section 4.5), jumps and determine the minimal martingale measure. The Black-Scholes formulae for Markov-modulated Brownian (B, S)-security markets (4.4) without, (Subsection 4.4.2), and with jumps, (Subsection 4.5.2), are derived.

Perfect hedging in a Markov-modulated Brownian and Brownian fractional (B, S)-security market, (without and with jumps), is not possible since we have an incomplete market. Following the idea proposed by Föllmer & Sondermann [13] and Föllmer & Schweizer [12] we look for the strategy which *locally minimizes the risk.* The residual risk processes are derived for all these schemes.

Variance swaps when the stochastic volatility is driven by Markov process are also studied (Section 5). An example of variance swaps where the stochastic volatility is driven by a two-state continuous Markov chain is discussed in Section 5.3.

A Feynmann-Kac's formula for the general Markov-modulated Process $(y_s(t), x_s(t))_{t \geq s}$ (Appendix A.1) and a formula for the option price $f_T(S_T)$ for the market when there is also a compound geometric Poisson process (Appendix A.2) are presented in the Appendix.

4.2 Literature review

Black and Scholes [2] obtained the option pricing formula for the Brownian market. Merton [23] extended the result to the case where the the stock returns are discontinuous. Cox and Ross [4] valued options for alternative stochastic processes. Oldfield, Rogalski and Jarrow [25] considered autoregressive jump process for common stock returns. Harrison and Pliska [18] introduced and studied arbitrage and completeness of Brownian market. Föllmer and Sondermann [13] introduced and studied locally minimizing risk strategies. Aase [1] obtained option pricing formula when the security price is a combination of an Itô process and a random point process. Hamilton [17] introduced Markov switching into the econometric mainstream. Föllmer and Schwiezer [12] studied hedging under incomplete information using the minimal martingale measure. Elliott and Föllmer's paper [10] studies an optional stochastic integrals which are the sum of a predictable stochastic integral of a martingale and an orthogonal martingale, and their applications in finance. Di Masi, Platen and Runggaldier [5] considered the hedging of options under discrete observations of assets with stochastic volatility in a discrete time framework. Di Masi, Kabanov and Runggaldier [6] obtained option pricing formula for stochastic volatility driven by a Markov chain in continuous time. Hofmann, Platen and Schweizer [19] studied option pricing under incompleteness and with stochastic volatility. Swishchuk [32] obtained an option pricing formula for a model with stochastic volatility driven by a semi-Markov process. Gray [14] combined GARCH effects with Markov switching. Griego and Swishchuk [15] obtained the Black-Scholes formula for a market in a Markov random environment. Elliott and Swishchuk [7] studied option pricing formulae and swaps for Markov-modulated Brownian and fractional Brownian Markets with jumps. The jumps in the dynamics of stock prices have been considered, in particular, by Merton [23] and Aase [1]. In the paper by Elliott, Chan and Siu, [8], the authors consider the option pricing problem when the risky underlying assets are driven by a Markov-modulated geometric Brownian motion. It was shown that the martingale measure pricing measure chosen by a regime switching Esscher transform is the minimal entropy martingale measure with respect to the relative entropy. Variance and volatility swaps for financial markets with stochastic volatility that follow Heston model have been studied in Swishchuk [29] and variance swaps for financial markets with stochastic volatilites with delay have been studied in Swishchuk [28].

4.3 Martingale characterization of Markov processes

Let x_t be a homogeneous continuous-time Markov process in a locally compact phase space of states X with infinitesimal operator Q and suppose $x_0 = x$.

Lemma 1. *Let $f \in Domain(Q)$. Then the process*

$$m_t^f := f(x_t) - \int_0^t Qf(x_s))ds \tag{4.5}$$

is a martingale with respect to the filtration $F_t^* := \sigma\{x_s; 0 \le s \le t\}$.

Proof. Since x_t is a Markov process, m_t^f is adapted to the filtration F_t^*. For $0 \le s \le t \le T$ we have

$$m_t^f - m_s^f = f(x_t) - f(x_s) - \int_s^t Qf(x_u)du. \tag{4.6}$$

We note that

$$E\big[f(x_t) \mid F_t^*\big] = E\big[f(x_t) \mid x_s\big]$$

since x_t is a Markov process. If $T_t f(x_s) := E_x[f(x_t) \mid x_s]$ (we note that $T_t f(x) = E[f(x_t) \mid x_0 = x] := E_x[f(x_t)]$) then

$$T_t f(x_s) = f(x_s) + \int_s^t E\big[Qf(x_u) \mid F_s^*\big]du = f(x_s) + \int_s^t T_s Qf(x_u)du. \tag{4.7}$$

Hence, taking into account equation (7), we obtain

$$\begin{aligned} E\big[m_t^f - m_s^f \mid F_s\big] &= E\big[f(x_t) \mid F_s\big] - f(x_s) - E\Big[\int_0^t Qf(x_u)du \,\Big|\, F_s\Big] \\ &= T_t f(x_s) - f(x_s) - \int_s^t T_u Qf(x_u)du \\ &= 0, \end{aligned} \tag{4.8}$$

and so m_t^f is an F_t-martingale. □

Let us calculate the quadratic variation of the martingale m_t^f.

Lemma 2. *Let Q be such that if $f \in Domain(Q)$, then $f^2 \in Domain(Q)$. The quadratic variation $< m_t^f >$ of the martingale m_t^f in* (4.5) *is equal to*

$$\langle m_t^f \rangle = \int_0^t \big[Qf^2(x_s) - 2f(x_s)Qf(x_s)\big]ds. \tag{4.9}$$

Proof. We note that

$$\begin{aligned} (m_t^f)^2 &= f^2(x_t) - 2f(x_t)\int_0^t Qf(x_s)ds + \Big(\int_0^t Qf(x_s)ds\Big)^2 \\ &= f^2(x_t) - 2m_t^f \int_0^t Qf(x_s)ds - \Big(\int_0^t Qf(x_s)ds\Big)^2. \end{aligned} \tag{4.10}$$

Furthermore,

$$\begin{aligned}
d\Big[2m_t^f\int_0^t Qf(x_s)ds+\Big(\int_0^t Qf(x_s)ds\Big)^2\Big]
&= 2\int_0^t Qf(x_s)ds dm_t^f + 2Qf(x_t)dtm_t^f \\
&\quad + 2Qf(x_t)dt\int_0^t Qf(x_s)ds \\
&= 2\Big(\int_0^t Qf(x_s)ds\Big)dm_t^f + 2f(x_t)Qf(x_t)dt \\
&\quad - 2Qf(x_t)dt\int_0^t Qf(x_s)ds + 2Qf(x_t)dt\int_0^t Qf(x_s)ds \\
&= 2f(x_t)Qf(x_t)dt + 2\Big(\int_0^t Qf(x_s)ds\Big)dm_t^f.
\end{aligned} \tag{4.11}$$

Hence, from equations (10) - (11) we have

$$\begin{aligned}
(m_t^f)^2 &= f^2(x_t) - 2f(x_t)Qf(x_t)dt - 2\Big(\int_0^t Qf(x_s)ds\Big)dm_t^f \\
&= f^2(x_t) - \int_0^t Qf^2(x_s)ds - 2\Big(\int_0^t Qf(x_s)ds\Big)dm_t^f \\
&\quad + \int_0^t \big[Qf^2(x_s) - 2f(x_s)Qf(x_s)\big]ds.
\end{aligned} \tag{4.12}$$

Since $f^2 \in Domain(Q)$, then $f^2(x_t) - \int_0^t Qf^2(x_s)ds$ is a martingale, and m_t^f is also martingale. Then, $2\int_0^t(\int_0^s Qf(x_u)du)dm_s^f$ is a martingale too. Therefore, $(m_t^f)^2 - \int_0^t[Qf^2(x_s) - 2f(x_s)Qf(x_s)]ds$ is a martingale, so

$$\langle m_t^f\rangle = \int_0^t \big[Qf^2(x_s) - 2f(x_s)Qf(x_s)\big]ds. \tag{4.13}$$

□

Lemma 3. *Suppose the following condition (Novikov's condition) is satisfied*

$$E^P \exp\Big\{\frac{1}{2}\int_0^t \big[Qf^2(x_s) - 2f(x_s)Qf(x_s)\big]ds\Big\} < +\infty, \quad \forall f^2 \in Domain(Q).$$

Then $E^P e_t^f = 1$, *where*

$$e_t^f := e^{m_t^f - \frac{1}{2}\langle m_t^f\rangle}, \tag{4.14}$$

and e_t^f *in (14) is a P-martingale (Doléans-Dade martingale).*

4.4 Pricing options for Markov-modulated security markets

4.4.1 Incompleteness of Markov-modulated Brownian security markets

A standard market is described by the following system of price processes

$$\begin{cases} dB_t = rB_t dt, \quad B_0 > 0, \\ dS_t = S_t(\mu dt + \sigma dw_t), \quad S_0 > 0. \end{cases} \tag{4.15}$$

Here B is the riskless asset, (bond), with an interest rate $r > 0$, and S is the risky asset, (stock), with appreciation rate μ and volatility σ. The process w is the standard one-dimensional Brownian motion.

Suppose now that the parameters $r \equiv r(x)$, $\mu \equiv \mu(x)$ and $\sigma \equiv \sigma(x)$ depend on some parameter x and these functions are continuous and bounded, $r(x) > 0$ and $\sigma(x) > 0$, $\forall x \in X$. Furthermore, assume that x varies as a Markov process x_t. The dynamics of B and S are:

$$\begin{cases} dB_t = r(x_t)B_t dt, \quad B_0 > 0, \\ dS_t = S_t(\mu(x_t)dt + \sigma(x_t)dw_t), \quad S_0 > 0, \end{cases} \tag{4.16}$$

The Markov process x_t is thus an additional source of randomness. The process (B_t, S_t) is called a *Markov modulated Brownian market.*

Let $\pi_t := (\beta_t, \gamma_t)$ be a portfolio (strategy) at time t, which is F_t-measurable, and $X_t^\pi := \beta B_t + \gamma_t S_t$ be a capital (or wealth process) at time $t, 0 \le t \le T$.

The following definitions can be found in Musiela and Rutkowski [24].

Definition 1. *Strategy π_t is called self-financing if $B_t d\beta_t + S_t d\gamma_t = 0$.*

Definition 2. *A self-financing strategy π generates an arbitrage opportunity if $P(X_0^\pi = 0) = 1$, and $P(V_T^\pi \ge 0) = 1$ and $P(V_T^\pi > 0) > 0$.*

Definition 3. *The market is arbitrage-free if there are no arbitrage opportunities in the class of self-financing strategies.*

Definition 4. *A European contingent claim V which settles at time T is an arbitrary F_T-measurable random variable.*

Definition 5. *A replicating strategy for the cintingent claim V, which settles at time T, is a self-financing startegy π such that $X_T^\pi = V$.*

Definition 6. *A claim V is attainable if it admits at least one replicating strategy.*

Definition 7. *The market is complete if every claim V is attainable. Otherwise, it is incomplete.*

Definition 8. *The measure P^* is called a martingale measure, if it is equivalent to P and such that the discounted capital $M_t := \frac{X_t}{B_t}$ is P^*-martingale, that is, $M_0 = E^{P^*}(M_T)$.*

We note that arbitrage-free market is complete if and only if there exists a unique martingale measure (Harrison and Pliska [18]).

Define the following process

$$\eta_t := e^{\int_0^t [(r(x_s)-\mu(x_s))/\sigma(x_s)]dw_s - \frac{1}{2}\int_0^t [(r(x_s)-\mu(x_s))/\sigma(x_s)]^2 ds} \tag{4.17}$$

and measure $\hat{P}$

$$\frac{d\hat{P}}{dP}\bigg|_{F_T} = \eta_T. \tag{4.18}$$

Theorem 1. *Suppose the Markov process x_t is independent of w_t and the following two conditions (Novikov's conditions) are satisfied:*

$$E^P\left\{\exp\frac{1}{2}\int_0^t [(r(x_s)-\mu(x_s))/\sigma(x_s)]^2 ds\right\} < +\infty \tag{4.19}$$

and

$$E^{\hat{P}}\exp\left\{\frac{1}{2}\int_0^t [Qf^2(x_s) - 2f(x_s)Qf(x_s)]ds\right\} < +\infty, \quad \forall f^2 \in Domain(Q), \tag{4.20}$$

where $\hat{P}$ is defined in (4.18).

Then the Markov-modulated security market is incomplete.

Proof. We shall prove that there are two distinct equivalent martingale measures, and hence, that the (B,S)-security market is incomplete.

From the Novikov's condition (4.19) it follows that $E^P\eta_T = 1$, where η_T is defined in (4.17). Also, from the Novikov's condition (4.20) it follows that $E^{\hat{P}}E_T^f = 1$, where E_T^f is defined in (4.14). We note, that by conditions (4.19) and (4.20) the processes η_t in (4.19) and E_t^f in (4.14) are P-martingale and $\hat{P}$-martingale, respectively. It is not difficult to see that $\hat{P}$ is a probability measure, since

$$\hat{P}(\Omega) = \int_\Omega d\hat{P} = \int_\Omega \eta_T dP = E^P\eta_T = 1.$$

By Girsanov's theorem the process

$$\hat{w}_t := w_t - \int_0^t \left[(r(x_s) - \mu(x_s))/\sigma(x_s) \right] ds$$

is a $\hat{P}$-Wiener process, where $\hat{P}$ is defined in (4.18).

Write

$$M_t := \frac{S_t}{B_t},$$

where B_t and S_t are defined in (4.16).

Then, by the Itô formula M_t satisfies the equation

$$M_t = M_0 + \int_0^t \frac{\sigma(x_u) S_u}{B_u} d\hat{w}_u, \tag{4.21}$$

so that the process M_t is a $\hat{P}$-martingale.

Define the measure $\tilde{P}$ through

$$\left. \frac{d\tilde{P}}{dP} \right|_{F_T} = \eta_T E_T^f, \tag{4.22}$$

where E_T^f are defined in (4.14) and η_T in (4.17). It is not difficult to see that $\tilde{P}$ is a probability measure. Indeed, from Lemma 3 and condition (4.20) it follows that

$$\tilde{P}(\Omega) = \int_\Omega d\tilde{P} = \int_\Omega \eta_T E_T^f dP = \int_\Omega E_T^f d\hat{P} = E^{\hat{P}} E_T^f = 1.$$

It is easy to see that M_t in (4.21) is also $\tilde{P}$-martingale, that follows from Lemma 13.10, p. 190 (see Elliott [11]). Therefore, we have two distinct equivalent martingale measures, namely, $\hat{P}$ and $\tilde{P}$ in (4.18) and (4.22), respectively. Hence, the Markov-modulated security market is incomplete.

4.4.2 The Black-Scholes formula for pricing options in a Markov-modulated Brownian market

Consider the price of a European contingent claim X at time T. If X is attainable, the price of X at time t is given by any equivalent martingale measure (e.g. $\tilde{P}$ or $\hat{P}$) and

$$C_t = B_t E^{\hat{P}} [X B_T^{-1} \mid F_t].$$

C_t is also called the no-arbitrage price.

We note that if X is not attainable, the risk-minimizing hedge price can be used.

Definition 9. *Two martingales are said to be strongly orthogonal if their product follows a martingale.*

Definition 10. *A martingale measure P^* for discounted capital is called a minimal martingale measure associated with P if any local P-martingale strongly orthogonal (under P) to each local martingale M remains a local martingale under P^*.*

Lemma 4. *The measure $\hat{P}$ in (4.18) is the minimal martingale measure associated with P.*

Proof. Suppose w_t and x_t are F_t-adapted. If N_t is an L^2-P local martingale, then by the Kunita-Watanabe representation (see Elliott [11], Elliott and Föllmer [10], Musiela and Rutkowski [24], Kallianpur and Karandikar [21])

$$N_t = N_0 + \int_0^t \beta_u dw_u + \int_0^t \beta'_u dm_u^f + z_t,$$

for any function $f \in Domain(Q)$, where $\langle w_t, z_t\rangle = \langle m_t^f, z_t\rangle = 0$. Let N be strongly orthogonal to $\int_0^t \sigma_u(x_u)dw_u$. Then we have

$$\begin{aligned}
0 = \Big\langle N, \int_0^t \sigma(x_u)dw_u\Big\rangle &= \Big\langle N_0 + \int_0^t \beta_u dw_u + \int_0^t \beta'_u dm_u^f + z_t, \int_0^t \sigma(x_u)dw_u\Big\rangle \\
&= \Big\langle N_0, \int_0^t \sigma(x_u)dw_u\Big\rangle + \Big\langle \int_0^t \beta_u dw_u, \int_0^t \sigma(x_u)dw_u\Big\rangle \\
&\quad + \Big\langle \int_0^t \beta'_u dm_t^f, \int_0^t \sigma(x_u)dw_u\Big\rangle + \Big\langle z_t, \int_0^t \sigma(x_u)dw_u\Big\rangle \\
&= \int_0^t \beta_u \sigma(x_u)du.
\end{aligned}$$

Hence, $\beta_u = 0$ a.e. for all $u \in [0,T]$, as $\sigma(x) > 0$. Therefore,

$$\begin{aligned}
d(N_t\eta_t) &= N_t d\eta_t + \eta_t dN_t + d\langle N, \eta\rangle_t \\
&= N_t d\eta_t + \eta_t dN_t \\
&\quad + d\Big\langle N_0 + \int \beta_u dw_u \\
&\quad + \int \beta'_u dm_u^f + z_t, \int \big[\big(r(x_u) - \mu(x_u)\big)/\sigma(x_u)\big] dw_u\Big\rangle_t \\
&= N_t d\eta_t + \eta_t dN_t + d\Big\langle \int \beta_u dw_u, \int \big[\big(r(x_u) - \mu(x_u)\big)/\sigma(x_u)\big] dw_u\Big\rangle_t \\
&= N_t d\eta_t + \eta_t dN_t + \beta_t\gamma_t dt \\
&= N_t d\eta_t + \eta_t dN_t,
\end{aligned}$$

where η_t is defined in (4.17), and

$$\gamma_t := \big(r(x_t) - \mu(x_t)\big)/\sigma(x_t).$$

This means that $N_t\eta_t$ is local P-martingale. Hence, N_t is a local $\hat{P}$-martingale, and, finally, $\hat{P}$ is the minimal martingale measure.

Theorem 2. *Let $X := f_T(S_T)$ be a European contingent claim settled at time T (not necessarily attainable). Then the risk-minimizing hedge price is*

$$C_t(x,S) = B_t E_x^{\hat{P}}\big[f_T(S_T)B_T^{-1} \mid F_t\big]. \tag{4.23}$$

Proof. By Lemma 4, $\hat{P}$ is the unique minimal equivalent martingale measure. We note that the backward Cauchy problem for $C_t(x,S)$ is given by

$$\begin{cases} \dfrac{\partial C}{\partial t} + L(x)C - r(x)C + QC = 0 \\ \qquad\qquad\qquad\qquad C_T(x,S) = f_T(S) \end{cases}$$

where $f(S)$ is a bounded continuous function on R_+ and $L(x)$ is the differential operator

$$L(x) \;=\; r(x)\cdot s\cdot \frac{d}{ds} + \frac{1}{2}\sigma^2(x)\cdot s^2\cdot \frac{d^2}{ds^2},$$

having the solution

$$C_t(x,S) \;=\; E_x^{\hat{P}}\Big[f(S_{T-t})\exp\Big(-\int_t^T r(x_t(\nu))d\nu\Big)\Big].$$

The theorem follows directly from Theorem 6 in the Appendix with $r(t,x,y) \equiv -r(x)$, for all $t \geq 0$ and $y \in R$. Thus, the risk-minimizing hedge price is given by (4.23).

Corollary 1. *In particular, for the European call options $f_T(y) = f(y) = (y-K)^+$, we have*

$$C_0(x,S) = \int C_{BS}\big((z/T)^{1/2}, T, S\big)F_T^x(dz), \tag{4.24}$$

where $C_{BS}(\hat{\sigma}, T, S)$, $S := S_0$, is the Black-Scholes price for the call option with volatility $\hat{\sigma}$, i. e.,

$$C_{BS}(\hat{\sigma}, T, S) = S\Phi(d_+) - Ke^{-rT}\Phi(d_-), \tag{4.25}$$

where

$$d_{\pm} = \left[\ln\frac{S}{K} + rT \pm \frac{\hat{\sigma}^2 T}{2}\right]/\hat{\sigma}\sqrt{T}, \tag{4.26}$$

and F_t^x is a distribution of a random variable

$$Z_t^x = \int_0^t \sigma^2(x_u)\,du. \tag{4.27}$$

Proof. For a standard European call option with the cost function $f_T(S) = (S_T - K)^+$, the option price $C_T(x,S)$ is defined by the formula

$$C_T(x,S) \;=\; E_x^{\hat{P}}\left[(S(T)-K)^+ \exp\Big(-\int_0^T r(x_\nu)d\nu\Big)\right]$$

where

$$S(T) = S_0 \exp\Big(\int_0^T r(x_s)ds\Big)\exp\Big(\int_0^T \sigma(x_s)dW_s - \frac{1}{2}\sigma^2(x_s)ds\Big)$$
$$S = S_0.$$

The formula for $C_T(x,S)$ is obtained from Theorem 2 by letting $f_T(S) = (S_T-K)^+$. The value $C_T(x,S)$ can be calculated in some cases more simply. For example, letting $r(x)=r$ for all $x\in X$, it follows from above that

$$C_T(x,S) \;=\; E_x^{\hat{P}}\big[\max(S_T-K,0)\big],$$

where

$$S(T) = S_0 e^{rT}\exp\Big(\int_0^T \sigma(x_s)dW_s - \frac{1}{2}\sigma^2(x_s)ds\Big).$$

We note that the function

$$C_t(x,S) \;=\; E_x^{\hat{P}}[f(S_{T-t})]$$

is the solution of the boundary value problem

$$\begin{cases} \dfrac{\partial C}{\partial t} + rS\dfrac{\partial C}{\partial S} + \dfrac{1}{2}\sigma^2(x)s^2\dfrac{\partial^2 C}{\partial s^2} - rC + QC = 0 \\ \qquad\qquad\qquad\qquad\qquad\qquad C_T(x,S) = f(S) \end{cases}$$

where $dS_t = rS_t dt + \sigma(x_t)S_t d\hat{w}(t),\ S_0 = S.$

Let F_T^x be the distribution of the random variable $Z_T^x \equiv \int_0^T \sigma^2(x_s)ds$. Then, from the above it follows that

$$C_0(x,S) = E_x^{\hat{P}}[f(S_T)] = \int\Big(\int f(y)y^{-1}\psi\Big(z, ln\frac{y}{s}+\frac{1}{2}z\Big)dy\Big)\,F_T^x(dz)$$

where $\psi(z,\nu) = (2\pi z)^{-1/2}\exp\left(\frac{\nu^2}{2z}\right)$.

In particular, for $f(s) \;=\; (s-K)^+$, we have for all $x\in E$,

$$C_0(x,S) = \int C_T^{BS}\Big(\Big(\frac{z}{T}\Big)^{1/2}, T, S\Big)\,F_T^x(dz)$$

where $C_T^{BS}(\sigma,T,S)$ is a Black-Scholes value for a European call option with volatility σ, expiration date T and interest rate r. □

Remark 3. Perfect hedging in Markov-modulated Brownian (B, S)-security market is not posssible since we have an incomplete market. Following the idea proposed by Föllmer and Sondermann [13] and Föllmer and Schweizer [12] we look for the strategy *locally minimizing the risk.* The strategy π^* is locally risk-minimizing, if for any *H-admissible* ($X_T = H$, where X_T is a capital at time T) strategy π and any t

$$R_t(\pi^*) \leq R_t(\pi),$$

where the *residual risk* is defined as follows

$$R_t(\pi) := E_x^{\hat{P}}\left([C_T(\pi) - C_t(\pi)]^2 / F_t\right),$$

and

$$C_t(\pi) := X_t(\pi) - \int_0^t \beta_u dS_u,$$

β_t is the number of stocks at time t, $E_x^{\hat{P}}$ is the expectation with respect to measure $\hat{P}$ conditionally $x_0 = x$.

It may be shown (see Swishchuk [32]) that the residual risk process can be expressed as

$$R_t(\pi^*) = E_x^{\hat{P}}\left(\int_t^T \left[Qu^2(r, S_r, x_r) - 2u(r, S_r, x_r)Qu(r, S_r, x_r)\right] dr \;\Big|\; F_t\right),$$

where the function u satisfies the following boundary value problem

$$\begin{cases} u_t(t, S, x) + rSu_S(t, S, x) + \frac{1}{2}\sigma^2(x) \cdot S^2 \cdot u_{SS}(t, S, x) + Qu(t, S, x) = 0 \\ u(T, S, x) = f(S). \end{cases}$$

In particular the residual risk at time $t = 0$ is equal to

$$R_0(\pi^*) = E_x^{\hat{P}}\left(\int_0^T \left[Qu^2(r, S_r, x_r) - 2u(r, S_r, x_r)Qu(r, S_r, x_r)\right] ds\right),$$

where the operator Q is an infinitesimal operator of the process x_t.

Remark 4. Let $X = \{1, 2\}$, and $\nu(t)$ be a counting jump process for x_t. Then the distribution F_t^x of random variable Z_T^x may be expressed in explicit form

$$\begin{aligned} Z_T^1 &= \int_0^T \left[\sigma^2(4.1)I(x_t = 1) + \sigma^2(4.2)I(x_t = 2)\right] dt \\ &= a \cdot T + b \cdot J_T, \end{aligned}$$

where

$$J_T = \int_0^T (-1)^{\nu(t)}\, dt$$

$$a = \frac{1}{2}(\sigma^2(4.1) + \sigma^2(4.2)),$$

$$b = \frac{1}{2}(\sigma^2(4.1) - \sigma^2(4.2)).$$

The formula for the distribution of J_T may be found in Di Masi et al [6] and contains modified first order Bessel functions.

4.5 Pricing options for Markov-modulated Brownian markets with jumps

4.5.1 Incompleteness of Markov-modulated Brownian (B, S)-security markets with jumps

Suppose we have a Markov-modulated Brownian (B, S)-security market (4.16) on the interval $[\tau_k, \tau_{k+1})$. Assume that at the moment τ_k we have the jump in S_t. That is,

$$S_{\tau_k} - S_{\tau_k -} = S_{\tau_k} u_k, \tag{4.28}$$

where u_k, $k \geq 1$, are independent identically distributed random variables with values in $(-1, +\infty)$ with distribution function $H(dy)$. The moments τ_k are the moments of jumps for the Poisson process N_t with intensity $\lambda > 0$. We suppose that τ_k, u_k, are independent on x_t and w_t, $\quad k \geq 1$.

Denote F_t the σ-algebra generated by the random variable $w_t, N_t,$ and $u_j \mathbf{1}_{\{j \leq N_t\}}$ for $j \geq 1$, where $\mathbf{1}_A = 1$, if $\omega \in A$, and $\mathbf{1}_A = 0$, if $\omega \notin A$.

It can be shown that w_t is a standard Brownian motion with respect to F_t. N_t is a process adapted to this filtration and $N_t - N_s$ is independent of the σ-algebra F_t for all $t > s$.

Taking into account (4.17) and (4.28) we obtain

$$S_t = S_0 \Big(\prod_{j=1}^{N_t} (1 + u_j) \Big) e^{\int_0^t [\mu(x_s)) - \sigma^2(x_s)/2] ds + \int_0^t [\sigma(x_s)] dw_s} \tag{4.29}$$

with the convention that $\prod_{j=1}^{0} = 1$. We note that S_t in (4.29) may be written down in the following form

$$S_t = S_0 e^{\int_0^t [\mu(x_s)) - \sigma^2(x_s)/2] ds + \int_0^t [\sigma(x_s)] dw_s + \int_0^t \int_{-1}^{+\infty} \ln(1+y) \nu(dy, ds)}, \tag{4.30}$$

where $\nu(A, t)$ is a random point measure equal to the number of jumps that the process N_t makes before time t with values in the Borel set $A \subset R$. With this, we have another source of randomness (besides the Brownian motion w_t and Markov process x_t) for the (B, S)-security market.

Theorem 3. *The (B,S)-security market with jumps consisting of the stock in (4.30) and bond in (4.16) is incomplete under the conditions in (4.19) and (4.20).*

Proof. Let

$$\eta_t^* := e^{\int_0^t [(r(x_s)-\mu(x_s))/\sigma(x_s)]dw_s - \frac{1}{2}\int_0^t [(r(x_s)-\mu(x_s))/\sigma(x_s)]^2 ds} \prod_{k=1}^{N_t} h(u_k), \quad (4.31)$$

where $h(y)$ is a such function that

$$\begin{cases} \int_R h(y)H(dy) = 1, \quad \text{and} \\ \int_R yh(y)H(dy) = 0, \end{cases} \quad (4.32)$$

where $H(dy)$ is a distribution on $(-1,+\infty)$, with respect to $(u_k;\ k \geq 1)$. We note, that $(\lambda, H(dy))$ is a (P, F_t)-local quadratic variation of the compound Poisson process $\sum_{k=1}^{N_t} u_k$ independent of w_t.

Let P^* be a measure such that

$$\left.\frac{dP^*}{dP}\right|_{F_T} = \eta_T^*, \quad (4.33)$$

where η_t^* is defined in (4.31). Than P^* is a probability measure by the same arguments as in Theorem 1 (we note that $\prod_{k=1}^{N_t} h(u_k)$ is independent of w_t and m_t^f). By Girsanov's theorem the process

$$w_t^* := w_t - \int_0^t \gamma_s ds,$$

is a P^*-Wiener process, where P^* is defined in (4.33), and

$$\gamma_s := \big(r(x_s) - \mu(x_s)\big)/\sigma(x_s).$$

Let

$$M_t^* := \frac{S_t}{B_t},$$

where S_t and B_t are defined in (4.30) and (4.16), respectively. Then, by Itô's formula M_t^* satisfies the equation

$$M_t^* = M_0^* + \int_0^t \frac{\sigma(x_u)S_u}{B_u} dw_u^*$$

and the process M_t^* is a P^*-martingale. Introduce the measure $\bar{P}$ via

$$\left.\frac{d\bar{P}}{dP}\right|_{F_T} = \eta_T^* e_T^f, \quad (4.34)$$

where e_T^f and η_T^* are defined in (4.14) and (4.31), respectively. We note, that $\bar{P}$ is a probability measure by the same arguments as in Theorem 1, Section 3.1 (we note that $\prod_{k=1}^{N_t} h(u_k)$ is independent of w_t and m_t^f). It is easy to see that M_t^* is also $\bar{P}$-martingale, that follows from Lemma 13.10, p. 190 (see Elliott [11]).

Therefore, we have two distinct equivalent martingale measure namely, $\bar{P}$ and P^* in (4.33) and (4.34), respectively.

Hence, the (B, S)-security market with jumps is incomplete and Theorem 3 is proved. □

4.5.2 Black-Scholes formula for pricing options in Markov-modulated Brownian (B, S)-security market with jumps

Using the reasonings in establishing (4.23)-(4.24) we obtain the following results for the Markov-modulated (B, S)-security market with jumps (see Theorem 7, Appendix).

Theorem 4. *If X is attainable, the price of X at time t is given, under any equivalent martingale measure (e.g. P^* or $\bar{P}$),*

$$C_t(x, S) = B_t E^{\bar{P}}\left[X B_T^{-1} \middle| F_t\right],$$

and C_t is called the no-arbitrage price.

If X is not attainable, we have the risk-minimizing hedge price, and P^* in (4.33) is the minimal martingale measure associated with P, which can be proved by the same arguments as in Lemma 4 using the fact that $\prod_{k=1}^{N_t} h(u_k)$ is independent of w_t and m_t^f). Then the risk-minimizing hedge price is

$$C_t(x, S) = B_t E^{P^*}\left[X B_T^{-1} \middle| F_t\right].$$

Corollary 2. *If $r(x) \equiv r$, $\forall x \in X$, then the price $C^T(x, S)$ of contingent claim $f_T(S_T)$ is calculated by the formula*

$$\begin{aligned} C^T(x, S) &= e^{-rT} \sum_{k=0}^{+\infty} \frac{\exp\{-\lambda T\}(\lambda T)^k}{k!} \\ &\times \int_{-1}^{+\infty} \cdots \int_{-1}^{+\infty} \Big(\int \Big(\int f(y) y^{-1} \\ &\times \psi\Big(z, \ln \frac{y}{S \prod_{i=1}^{k}(1 + y_i)} + rT + 2^{-1} z\Big) dy\Big) F_T^x(dz)\Big) \\ &\times H^*(dy_1) \times \ldots \times H^*(dy_k), \end{aligned} \tag{4.35}$$

where $H^*(dy) = h(y)H(dy)$, and $\psi(z, v) := (2\pi z)^{-2^{-1}} \exp\{-\frac{v^2}{2z}\}$.

Proof. This follows from the representation of S_t, formulae (A7)-(A10) (see Appendix) and iterations on function $f_T(S_T)$, taking into account a distribution of Z_T^x. □

We note that the function $C_T(x,S) = E^{P^*}[f_T(S_{T-t})]$ is the solution of the Cauchy problem:

$$\begin{cases} \dfrac{\partial C}{\partial t} + rS\dfrac{\partial C}{\partial S} + 2^{-1}\sigma^2(x)S^2 \times \dfrac{\partial^2 C}{\partial S^2} - rC \\ \qquad + \lambda \displaystyle\int_{-1}^{+\infty} (C(t,S(1+y)) - C(t,x))h(y)H(dy) + QC = 0, \\ C(T,S) = f_T(S). \end{cases}$$

In the case of $X := f_T(S_T) = (S_T - K)^+$, where K is a strike price, inserting the function $f_T(S_T) = (S_T - K)^+$ in the expression above (see Corollary 2) we obtain the following result.

Corollary 3. *Let F_T^x be a distribution of random variable $Z_T^x := \int_0^T \sigma^2(x(s))$ ds. Also, let $f_T(S_T) = (S_T - K)^+$, and $r(x) \equiv r$. Then from Theorem 4 and formula (4.35), where $C_T^{BS}(\sigma, T, S)$ is the Black-Scholes value for European call option it follows that the price $C_T(x,S)$ of contingent claim has the form*

$$\begin{aligned} C_0(x,S) = &\sum_{k=0}^{+\infty} \frac{\exp\{-\lambda T\}(\lambda T)^k}{k!} \\ &\times \int_{-1}^{+\infty} \dots \int_{-1}^{+\infty} \int C_{BS}\Big(\Big(\frac{z}{T}\Big)^{2^{-1}}, T, S\prod_{i=1}^{k}(1+y_i)\Big) F_T^x(dz) \\ &\times H^*(dy_1) \times \dots \times H^*(dy_k), \end{aligned} \tag{4.36}$$

where function $C_{BS}(\hat{\sigma}, T, S)$ is a Black-Scholes value for European call option (see (4.25)-(4.26)), F_t^x is a distribution of a random variable (see (4.27))

$$Z_t^x = \int_0^t \sigma^2(x_r)\, dr,$$

and $H^*(dy) := h(y)H(dy)$, where $h(y)$ is defined in (4.32).

Remark 5. Perfect hedging in Markov-modulated Brownian (B,S)-security market with jumps is not posssible since we have an incomplete market. We look for the *locally minimizing the risk* strategy.

The residual risk process (see Remark 3) is expressed in the following way

$$R_t(\pi^*) = E_x^{P^*}\Big(\int_t^T [Qu^2(r,S_r,x_r) - 2u(r,S_r,x_r)Qu(r,S_r,x_r)]\,dr \,\Big|\, F_t\Big),$$

where the function u satisfies the following boundary value problem

$$\begin{cases} u_t(t,S,x) + rSu_S(t,S,x) + \frac{1}{2}\sigma^2(x)\cdot S^2\cdot u_{SS}(t,S,x) \\ \qquad + \lambda \int_{-1}^{+\infty} [u(t,S(1+v),x) - u(t,S,x)]H^*(dv) \\ \qquad - ru + Qu(t,S,x) = 0 \\ u(T,S,x) = f(S). \end{cases}$$

In particular the residual risk at the moment $t = 0$ is equal to

$$R_0(\pi^*) = E_x^{P^*}\Big(\int_0^T [Qu^2(r,S_r,x_r) - 2u(r,S_r,x_r)Qu(r,S_r,x_r)]\,ds\Big),$$

where the operator Q is infinitesimal operator of the process x_t.

4.6 Pricing of Variancev swaps for stochastic volatility driven by Markov process

4.6.1 Stochastic volatility driven by Markov process

Let x_t be a Markov process in measurable phase space X with generator Q. The stock price S_t satisfies the stochastic differential eqution

$$dS_t = S_t\big(r(x_t)dt + \sigma(x_t)dw_t)\big)$$

with the volatility $\sigma := \sigma(x_t)$ depending on the process x_t, which is independent of the standard Wiener process w_t.

A variance swap is a forward contract on an annualized variance, the square of the realized volatility. Its payoff at expiration is equal to

$$N(\sigma_R^2(x) - K_{var}),$$

where $\sigma_R^2(x)$ is the realized stock variance (quoted in annual terms) over the life of the contract,

$$\sigma_R^2(x) := \frac{1}{T}\int_0^T \sigma^2(x_s)ds,$$

K_{var} is the delivery price for variance, and N is the notional amount of the swap in dollars per annualized volatility point squared. The holder of a variance swap at expiration receives N dollars for every point by which the stock's realized variance $\sigma_R^2(x)$ has exceeded the variance delivery price K_{var}.

4.6.2 Pricing of variance swaps for stochastic volatility driven by Markov process

Pricing a variance forward contract or swap is no different from valuing any other derivative security. The value of a forward contract F on future realized variance with strike price K_{var} is the expected present value of the future payoff in the risk-neutral world. That is,

$$P(x) = E\{e^{-rT}(\sigma_R^2(x) - K_{var})\},$$

where r is the risk-free discount rate corresponding to the expiration date T, and E denotes the expectation.

Let us show how we can calculate $EV(x)$, where $V(x) := \sigma_R^2(x)$. For this we need to calculate $E\sigma^2(x_t)$.

We note (see Section 2, Lemma 1) that for $\sigma(x) \in Domain(Q)$ the following process

$$m_t^\sigma := \sigma(x_t) - \int_0^t Q\sigma(x_s))ds$$

is a zero-mean martingale with respect to the filtration $F_t := \sigma\{x_s; 0 \le s \le t\}$.

The quadratic variation of the martingale m_t^f by Lemma 2 is equal to

$$\langle m_t^\sigma \rangle = \int_0^t \left[Q\sigma^2(x_s) - 2\sigma(x_s)Q\sigma(x_s)\right]ds, \quad \sigma^2(x) \in Domain(Q). \tag{4.37}$$

Since $\sigma(x_s)$ satisfies the stochastic differential equation

$$d\sigma(x_t) = Q\sigma(x_t)dt + dm_t^\sigma$$

we obtain from Itô's formula (see Elliott and Kopp [9]) the stochastic differential equation for $\sigma^2(x_t)$ given by

$$d\sigma^2(x_t) = 2\sigma(x_t)dm_t^\sigma + 2\sigma(x_s)Q\sigma(x_s)dt + d\langle m_t^\sigma \rangle, \tag{4.38}$$

where $\langle m_t^\sigma \rangle$ is defined in (4.37). Substituting (4.37) into (4.38) and taking the expectation of both parts in (4.38) we have

$$E\sigma^2(x_t) = \sigma^2(x) + \int_0^t QE\sigma^2(x_s)ds.$$

Solving the above equation we get

$$E\sigma^2(x_t) = e^{tQ}\sigma^2(x).$$

Finally, we obtain

$$EV(x) = \frac{1}{T}\int_0^T e^{tQ}\sigma^2(x)dt.$$

We therefore obtain the following result.

Theorem 5. *The value of a variance swap for Markov stochastic volatility* $\sigma(x_t)$ *is*

$$P(x) = e^{-rT}\Big(\frac{1}{T}\int_0^T e^{tQ}\sigma^2(x)dt - K_{var}\Big). \tag{4.39}$$

4.6.3 Example of variance swap for stochastic volatility driven by two-state continuous Markov chain

Let Q be a generator of two-state continuous time Markov chain, i.e.,

$$Q = \begin{pmatrix} q_{11} & q_{12} \\ q_{21} & q_{22} \end{pmatrix}$$

and

$$P(t) = \begin{pmatrix} p_{11}(t) & p_{12}(t) \\ p_{21}(t) & p_{22}(t) \end{pmatrix}$$

be a Markov transition function. Thus,

$$P(t) = e^{tQ}.$$

In this case, the variance takes two values: $\sigma^2(4.1)$ and $\sigma^2(4.2)$.

From formula (4.39) it follows that the value of a variance swap in this case is equal to

$$P(i) = e^{-rT}\Big(\frac{1}{T}\int_0^T \big[p_{i1}(s)\sigma^2(4.1) + p_{i2}(s)\sigma^2(4.2)\big]ds - K_{var}\Big) \tag{4.40}$$

for $i = 1, 2$.

It is apparent that the value of varinace swap depends on the initial state of Markov chain.

We note, that if a Markov chain is stationary with ergodic distribution (p_1, p_2), then the value of variance swap is

$$P = p_1 P(4.1) + p_2 P(4.2),$$

where $P(i)$, $i = 1, 2$, are defined in (4.40).

A Some auxiliary results

A.1 A Feynmann-Kac formula for the Markov-modulated process $(y_s(t), x_s(t))_{t\geq s}$

Let $\{x_s(t), t \geq s\}$ be a Markov process with state space X and infinitesimal matrix Q and with $x_s(s) = x \in X$, and let $\{y_s(t), t \geq s\}$ be the following process given by the equation

$$y_s(t) = y + \int_s^t \mu\big(\nu, x_s(\nu), y_s(\nu)\big)d\nu + \int_s^t \sigma\big(\nu, x_s(\nu), y_s(\nu)\big)dW(\nu)$$

where $y_s(s) = y$.

Also, let L_t be the differential operator

$$L_t = \mu(t,x,y)\frac{d}{dy} + \frac{1}{2}\sigma^2(t,x,y)\frac{d^2}{dy^2}$$

where the functions μ and σ are real-valued continuous and satisfy a Lipschitz condition.

The following theorem presents a Feynman-Kac formula for the Markov-modulated process $(y_s(t), x_s(t))_{t\geq s}$. See also Griego and Swishchuk, [15].

Theorem 6. *Let $r(t,x,y)$ be a bounded continuous function and consider a backward Cauchy problem for the function $u(t,x,y)$:*

$$\begin{aligned} \frac{\partial u}{\partial t} + L_t u + r(t,x,y)u + Qu = 0 \\ u(T,x,y) = \varphi(x,y) \end{aligned} \tag{A1}$$

where φ is a bounded continuous function $X \times R$. Then the Cauchy (A1) problem has the solution

$$u(t,x,y) = E_{t,x,y}\Big[\varphi\big(x_t(T), y_t(T)\big) \cdot \exp\Big(\int_t^T r\big(\nu, x_t(\nu), y_t(\nu)\big)d\nu\Big)\Big]. \tag{A2}$$

Here, $E_{t,x,y}$ is the integral with respect to the measure $P_{t,x,y}(\cdot) = P(\cdot|(x_t(t), y_t(t)) = (x,y))$.

Proof. Let $0 \leq s < t \leq T$ and consider the process

$$\zeta(t) \equiv u\big(t, x_s(t), y_s(t)\big) \cdot \exp\Big(\int_s^t r\big(\nu, x_s(\nu), y_s(\nu)\big)d\nu\Big). \tag{A3}$$

We note that $\zeta(t)$ is an F_s^t-martingale where $F_s^t \equiv \sigma(W(\nu), x_s(\nu) : s \leq \nu \leq t)$ with respect to $P_{s,x,y}$. We note that $F_t := F_0^t$.

Thus, we have

$$E_{s,x,y}\big[\zeta(u) - \zeta(t) \mid F_s^t\big] = 0. \tag{A4}$$

From (A3) and (A4) it follows that

$$E_{s,x,y}\big[\zeta(T)\big] = E_{s,x,y}\big[\zeta(s)\big].$$

But, taking into account (A1) we have

$$\begin{aligned} E_{s,x,y}\big[\zeta(T)\big] &= E_{s,x,y}\Big[u\big(T, x_s(T), y_s(T)\big) \cdot \exp\Big(\int_s^t r(\nu, x_s(\nu), y_s(\nu))d\nu\Big)\Big] \\ &= E_{s,x,y}\Big[\varphi\big(x_s(T), y_s(T)\big) \cdot \exp\Big(\int_s^t r(\nu, x_s(\nu), y_s(\nu))d\nu\Big)\Big] \end{aligned} \tag{A5}$$

and

$$E_{s,x,y}\big[\zeta(s)\big] = E_{s,x,y}\big[u\big(s, x_s(s), y_s(s)\big)\big] = u(s, x, y). \tag{A6}$$

Hence, from (A5) and (A6) we can conclude that

$$u(s, x, y) = E_{s,x,y}\Big[\varphi\big(x_s(T), y_s(T)\big) \cdot \exp\Big(\int_s^T r\big(\nu, x_s(\nu), y_s(\nu)\big)d\nu\Big)\Big]. \tag{A7}$$

Remark 6. Representation (A2) follows from (A7) with $s = t$. Theorem 6 is proved. □

A.2 Formula for the option price $f_T(S_T)$ for the market combined Markov-modulated (B, S)-security market and compound geometric Poisson process (see Section 4.4.2)

Theorem 7. *The price $C_0(x, S)$ of contingent claim $f_T(S_T)$ at time zero with expiry date T has the form*

$$C_0(x, s) = E^{P^*}\Big[f_T(S_T)\exp\Big\{-\int_0^T r(x(s))ds\Big\}\Big],$$

where P^* is risk-neutral measure in (4.33).

Proof. From Itô's formula it follows that S_t under the risk-neutral world is the solution of equation

$$dS_t = r\big(x(s)\big)S_t dt + \sigma\big(x(t)\big)S_t dw_t^* + S_t \int_{-1}^{+\infty} y\nu(dt, dy),$$

and $\nu(dt, dy)$ is a random measure, which equals the number of jumps of the Poisson process $N(t)$ with values in dy up to the moment dt. Hence, $(\lambda, H(dy))$ (see Section 4.1) is a local characteristic of measure $\nu(dt, dy)$ and $\tilde{\nu}(dt, dy) := \nu(dt, dy) - \lambda H(dy)$ is a local martingale.

We note that the following Cauchy problem

$$\begin{aligned} &\frac{\partial C}{\partial t} + r(x)S\frac{\partial C}{\partial S} + 2^{-1}\sigma^2(x)S^2\frac{\partial^2 C}{\partial S^2} - r(x)C \\ &\quad + \lambda\int_{-1}^{+\infty}(C(t, S(1+y)) - C(t, x))h(y)H(dy) + QC = 0, \\ &C_T(x, S) = f_T(S), \end{aligned}$$

has the solution

$$C_t(,x,S) = E^{P^*}\Big[f_T(S_{T-t}\exp\Big\{-\int_t^T r(x(s))ds\Big\}\Big],$$

which follows from the Black-Scholes equation for the Markov-modulated (B,S)-market (see Section 4.1) and Theorem 6, Appendix A.1. □

References

1. Aase, K. (1988). "Contingent claims valuation when the security price is a combination of an Itô process and random point process." *Stochastic Processes & Their Applications*, 28: 185–220.
2. Black, F. and M. Scholes (1973). "The pricing of options and corporate liabilities", *J. Polit. Economy*, May/June: 637–657.
3. Chung, K. and R. Williams (1983). *Introduction to Stochastic Integration*, Birkhauser, Boston.
4. Cox, J. and R. Ross (1976). "Valuation of options for alternative stochastic processes", *Joutrnal of Financial Economics.*
5. Di Masi, G. B., Platen, E. and W. Runggaldier (1994). "Hedging of options under discrete observation on assets with stochastic volatility."
6. Di Masi, G. B., Kabanov, Yu. M. and W. J. Runggaldier (1994). "Hedging of options on stock under mean-square criterion and Markov volatilities", *Theory Probability and its Applications*, 39(1): 211–222.
7. Elliott, R. and A. Swishchuk (2004). "Pricing Options and Variance Swaps in Markov-modulated Brownian and fractional Brownian Markets", *RJE 2005 Conference*, July 24–27, 2005, U of C, Calgary, AB, Canada (https://www.math.ucalgary.ca/~aswish/elliottswpaper1.pdf)
8. Elliott, R. J., Chan, L. L. and T. K. Siu (2005). "Option pricing and Esscher transform under regime switching". *Annals of Finance*, 1(4): 423-432.
9. Elliott, R. and E. Kopp (1999). *Mathematics of Financial Markets*, Springer.
10. Elliott, R. and H. Föllmer (1991). "Orthogonal martingale representation". *Technical Report 91.18*, Statistics Centre, University of Alberta, Edmonton, Canada, 14 (web page: http://www.stat.ualberta.ca/stats_centre/tech.htm)
11. Elliott, R. (1982). *Stochastic Calculus and Applications.* Springer-Verlag, New York.
12. Föllmer, H. and M. Schweizer (1991). "Hedging of contingent claims under incomplete information." *Applied Stochastic Analysis.* New-York-London: Gordon and Beach, 389–414.
13. Föllmer, H. and D. Sondermann (1986) "Hedging of nonredundant contingent claim." *Contributions to Mathematical Economics.*/Ed. by W. Hilenbrandt and A. Mas-Colell. Amsterdam-New-York: North-Holland, 205–223.
14. Gray, S. (1996). "Modelling the conditional distribution of interest rates as a regime-switching process". *Journal of Financial Econometrics*, 42: 27–62.

15. Griego, R. and A. Swishchuk (2000). "Black-Scholes Formula for a Market in a Markov Environment". *Theory of Probability and Mathematical Statistics* 62: 9–18.
16. Guo, X. (2001). "An explicit solution to an optimal stopping to an optimal stopping problem with regime switching". *Journal of Applied Probability*, 38: 464–481.
17. Hamilton, J. (1989). "Rational-expectations econometric analysis of changes in regime". *Journal of Economic Dynamics and Control*, 12: 385–423.
18. Harrison, J. and S. Pliska (1981). "Martingales, stochastic integrals and continuous trading". *Stochastic Processes and Applications*, 11 (3): 215–260.
19. Hofmann, N., Platen, E. and M. Schweizer (1994). "Options pricing under incompleteness and stochastic volatility". *Mathematical Finance*, 1996.
20. Hu, Y. and B. Øksendal (1999). "Fractional white noise calculus and applications to finance." *Preprint*, University of Oslo, Oslo, Norway.
21. Kallianpur, G. and R. Karandikar (2000). *Introduction to Option Pricing Theory*, Birkhauser.
22. Los, C. and J. Karupiah (1997). "Wavelet multiresolution analysis of high frequency Asian exchange rates". *Working Paper*, Department of Finance, Kent State University, Ohio, USA.
23. Merton, R. (1973). "Theory of rational option pricing". *Bell Journal of Economics and Management Science*, 4 (Spring): 141–183.
24. Musiela, M. and M. Rutkowski (1998). *Martingale Methods in Financial Modelling*, Springer-Verlag.
25. Oldfield, R. and R. Jarrow (1977). "Autoregressive jump process for common stock return". *Journal of Financial Economics.*
26. Rogers, L. C. G. (1997). "Arbitrage with fractional Brownian motion". *Mathematical Finance*, 7: 95–105
27. Shiryaev, A. (1998). "On arbitrage and replication for fractal models: In A. Sulem (ed.)". *Workshop on Mathematical Finance*, INRIA, Paris.
28. Swishchuk, A. (2005). "Modeling and pricing of variance swaps for stochastic volatilities with delay". *WILMOTT Magazine*, (September): 63–73.
29. Swishchuk, A. (2004). "Modeling of variance and volatility swaps for financial markets with stochastic volatilities." *WILMOTT magazine*, (September): 64–72.
30. Swishchuk, A. (2000). *Random Evolutions and Their Applications. New Trends.* Kluwer Academic Publishers, Dordrecht.
31. Swishchuk, A., Zhuravitskiy, D. and A. Kalemanova (2000). "Analogue of Black-Scholes formula for option prices of (B, S, X)-security markets with jumps." *Ukrainian Mathematical Journal*, 52 (3).
32. Swishchuk, A. (1995). "Hedging of options under mean-square criterion and with semi-Markov volatility". *Ukrainian Mathematical Journal*, 47, No. 7.

5

Smoothed Parameter Estimation for a Hidden Markov Model of Credit Quality

Małgorzata W. Korolkiewicz[1] and Robert J. Elliott[2]

[1] School of Mathematics and Statistics
University of South Australia
malgorzata.korolkiewicz@unisa.edu.au

[2] RBC Financial Group Professor of Finance
Haskayne School of Business
University of Calgary
relliott@ucalgary.ca

Summary. We consider a hidden Markov model of credit quality. We assume that the credit rating evolution can be described by a Markov chain but that we do not observe this Markov chain directly. Rather, it is hidden in "noisy" observations represented by the posted credit ratings. The model is formulated in discrete time with a Markov chain observed in martingale noise. We derive smoothed estimates for the state of the Markov chain governing the evolution of the credit rating process and the parameters of the model.

Key words: Hidden Markov model, smoothing, credit quality

5.1 Introduction

Spectacular growth in the market for credit derivatives in recent years has highlighted the importance of understanding credit quality. Credit ratings published in a timely manner by rating agencies are an invaluable source of credit risk information and Markov chain models have been used to describe their dynamics. The pioneering work in the direction of using Markov chain models, not only to describe the dynamics of a firm's credit rating but also to value credit derivatives, was done by Jarrow and Turnbull [6], and Jarrow, Lando and Turnbull [7].

Markov-type models assume that the credit rating process has no memory of its prior behaviour, i.e. that prior rating changes should have no predictive power for the direction of future rating changes. However, there exist empirical studies that suggest the contrary – the credit rating process seems to have memory. Two empirical studies of Moody's ratings, Carty and Fons [1], and Carty and Lieberman [2], found in particular that a firm upgraded (downgraded) was more likely to be subsequently upgraded (downgraded). More recently, Lando and Skodeberg [9] reported evidence of non-Markov effects for downgrades in a data set of Standard and Poor's ratings. We therefore consider a hidden Markov model of credit quality, where we assume that the credit rating evolution can be described by a Markov chain but we do not observe this Markov chain directly. Rather, it is hidden in "noisy" observations represented by the posted credit ratings.

Hidden Markov models, when Markov chains are observed in Gaussian noise, have been subject to extensive studies. See for example the book by Elliott, Aggoun and Moore [5] and references contained therein. Here we consider a discrete time model with a Markov chain observed in martingale noise. We derive smoothed estimates for the state of the Markov chain governing the evolution of the credit rating process and the parameters of the model.

The paper is organised as follows. Section 5.2 gives the dynamics of the Markov chain and observations. The reference probability measure is introduced in Section 5.3 and the forward filter in Section 5.4. Forward estimates for the processes needed to estimate the parameters of the model are obtained in Section 5.5. Finally, smoothed estimates and updating formulae are derived in Section 5.6.

5.2 Dynamics of the Markov chain and observations

We suppose that the signal process, the "true" credit quality, is a Markov chain which we do not observe directly. Rather, it is hidden in noisy observations represented by posted credit ratings.

Formally, a discrete-time, finite-state, time homogeneous Markov chain is a stochastic process $\{X_k\}$ with the state space $S = \{1, 2, \ldots, N\}$ and a transition matrix $A = (a_{ji})_{1 \le i,j \le N}$. Without loss of generality we can assume that $S = \{e_1, e_2, \ldots, e_N\}$, where $e_i = (0, \ldots, 0, 1, 0, \ldots, 0)^T \in \mathbb{R}^N$. Suppose that X is defined on the probability space $(\Omega, \mathcal{F}, P)$ and write $a_{ji} = P(X_{k+1} = e_j | \mathcal{F}_k) = P(X_{k+1} = e_j | X_k = e_i)$. Write $\mathcal{F}_k = \sigma\{X_0, X_1, \ldots, X_k\}$ for the σ-*field* containing all the information about the process X up to and including time k. Then, as shown in [5] and [8], $E[X_{k+1}|X_k] = AX_k$ and the *semi-martingale representation* of the chain X is

$$X_{k+1} = AX_k + V_{k+1}, \quad k = 0, 1, \ldots,$$

where V_{k+1} is a martingale increment with $E[V_{k+1}|\mathcal{F}_k] = 0 \in \mathbb{R}^N$.

Suppose we do not observe X directly. Rather, we observe a process Y such that

$$Y_k = c(X_k, \omega_k), \quad k = 0, 1, \ldots,$$

where c is a function with values in a finite set and $\{\omega_k\}$ is a sequence of independent identically distributed (IID) random variables independent of X. Suppose the range of c consists of M points which are identified with unit vectors $\{f_1, f_2, \ldots, f_M\}$, $f_j = (0, \ldots, 0, 1, 0, \ldots, 0) \in \mathbb{R}^M$.

Write $c_{ji} = P(Y_k = f_j | X_k = e_i)$, $1 \le i \le N$, $1 \le j \le M$. Then, $E[Y_k | X_k] = CX_k$, where $C = (c_{ji})_{1 \le i,j \le M}$ with $c_{ji} \ge 0$ and $\sum_{j=1}^{M} c_{ji} = 1$. Also, the semimartingale representation of the process Y is

$$Y_k = CX_k + W_k, \; k = 0, 1, \ldots,$$

where W is a martingale increment with $E[W_k | \mathcal{G}_{k-1} \vee \{X_k\}] = 0 \in \mathbb{R}^M$. Note that we are assuming zero delay between X_k and its observation Y_k.

5.3 Reference probability

Consider a probability measure $\bar{P}$ on $(\Omega, \mathcal{F})$ such that under $\bar{P}$, X is still a Markov chain with transition matrix A but $\{Y_k\}$ is a sequence of IID uniform variables independent of X. Suppose $C = (c_{ji})$, $1 \le i \le N$, $1 \le j \le M$, is a matrix with $c_{ji} \ge 0$, and $\sum_{j=1}^{M} c_{ji} = 1$.

Define $\bar{\lambda}_l = M \sum_{j=1}^{M} \langle CX_l, f_j \rangle \langle Y_l, f_j \rangle$ and $\bar{\Lambda}_k = \prod_{l=1}^{k} \bar{\lambda}_l$. Given filtrations

$$\begin{aligned} \mathcal{F}_k &= \sigma\{X_0, X_1, \ldots, X_k\}, \\ \mathcal{Y}_k &= \sigma\{Y_0, Y_1, \ldots, Y_k\} \\ \text{and} \quad \mathcal{G}_k &= \sigma\{X_0, \ldots, X_k, Y_0, \ldots, Y_k\}, \end{aligned}$$

define a new probability measure P by putting $\frac{dP}{d\bar{P}}\big|_{\mathcal{G}_k} = \bar{\Lambda}_k$. Then, as shown in [8], under P, X remains a Markov chain with transition matrix A and $P(Y_k = f_j | X_k = e_i) = c_{ji}$.

Note. P represents the "real world" probability measure. However, measure $\bar{P}$ is easier to work with since under $\bar{P}$, $\{Y_k\}$ is IID uniform and independent of X.

5.4 Recursive filter

Suppose we observe $Y_0, \ldots, Y_k$, and we wish to estimate $X_0, \ldots, X_k$. The best (mean-square) estimate of X_k given $\mathcal{Y}_k = \sigma\{Y_0, \ldots, Y_k\}$ is $E[X_k | \mathcal{Y}_k] \in \mathbb{R}^N$. However, $\bar{P}$ is a much easier measure under which to work. Using Bayes' theorem, we have

$$E[X_k \mid \mathcal{Y}_k] = \frac{\bar{E}[\bar{\Lambda}_k X_k \mid \mathcal{Y}_k]}{\bar{E}[\bar{\Lambda}_k \mid \mathcal{Y}_k]}.$$

Write $q_k := \bar{E}[\bar{\Lambda}_k X_k | \mathcal{Y}_k] \in \mathbb{R}^N$; q_k is then an unnormalized conditional expectation of X_k given the observations $\mathcal{Y}_k$. The dynamics of q_k are as follows:

$$\begin{aligned}
q_{k+1} &= \bar{E}[\bar{\Lambda}_{k+1} X_{k+1} \mid \mathcal{Y}_{k+1}] \\
&= \bar{E}\Big[\bar{\Lambda}_k (M \sum_{j=1}^{M} \langle C X_{k+1}, f_j\rangle \langle Y_{k+1}, f_j\rangle) X_{k+1} \;\Big|\; \mathcal{Y}_{k+1}\Big] \\
&= \sum_{i=1}^{N} \bar{E}\Big[\bar{\Lambda}_k (M \sum_{j=1}^{M} c_{ji} \langle Y_{k+1}, f_j\rangle) \langle X_{k+1}, e_i\rangle e_i \;\Big|\; \mathcal{Y}_{k+1}\Big] \\
&= \sum_{i=1}^{N} \bar{E}\Big[\bar{\Lambda}_k \langle X_{k+1}, e_i\rangle \;\Big|\; \mathcal{Y}_{k+1}\Big] \Big(M \sum_{j=1}^{M} c_{ji} \langle Y_{k+1}, f_j\rangle\Big) e_i \\
&= \sum_{i=1}^{N} \langle \bar{E}[\bar{\Lambda}_k (A X_k + V_{k+1}) \mid \mathcal{Y}_{k+1}], e_i\rangle \Big(M \sum_{j=1}^{M} c_{ji} \langle Y_{k+1}, f_j\rangle\Big) e_i \\
&= \sum_{i=1}^{N} \langle A \bar{E}[\bar{\Lambda}_k X_k \mid \mathcal{Y}_k], e_i\rangle \Big(M \sum_{j=1}^{M} c_{ji} \langle Y_{k+1}, f_j\rangle\Big) e_i \\
&= \sum_{i=1}^{N} \langle A q_k, e_i\rangle \Big(M \sum_{j=1}^{M} c_{ji} \langle Y_{k+1}, f_j\rangle\Big) e_i \\
&= B(Y_{k+1}) A q_k,
\end{aligned}$$

where $B(Y_{k+1})$ is a diagonal matrix with entries $M \sum_{j=1}^{M} c_{ji} \langle Y_{k+1}, f_j\rangle$.

Remark 1. Note that $\bar{E}[\bar{\Lambda}_k \mid \mathcal{Y}_k] = \langle q_k, \mathbf{1}\rangle$, where $\mathbf{1} = (1, \ldots, 1)' \in \mathbb{R}^N$.

5.5 Parameter estimates

To estimate parameters of the model, matrices A and C, we need estimates of the following processes:

$$J_k^{ij} = \sum_{n=1}^{k} \langle X_{n-1}, e_i\rangle \langle X_n, e_j\rangle, \quad 1 \le i, j \le N,$$

$$O_k^{i} = \sum_{n=1}^{k} \langle X_{n-1}, e_i\rangle, \quad 1 \le i \le N,$$

$$T_k^{ij} = \sum_{n=0}^{k} \langle X_n, e_i\rangle \langle Y_n, f_j\rangle, \quad 1 \le i \le N, \quad 1 \le j \le M.$$

The above processes are interpreted as follows:

J_k^{ij} – the number of jumps of X from state e_i to state e_j up to time k.

O_k^i – the amount of time the chain has spent in state e_i up to time $k-1$.

T_k^{ij} – the amount of time process X has spent in state e_i when process Y was in state f_j up to time k.

Remark 2. Note that $\sum_{j=1}^{N} J_k^{ij} = O_k^i$ and $\sum_{j=1}^{M} T_k^{ij} = O_{k+1}^i$.

Consider first the jump process $\{J_k^{ij}\}$. We wish to estimate J_k^{ij} given the observations $Y_0, \dots, Y_k$. Using Bayes' theorem, the best (mean-square) estimate is

$$E\left[J_k^{ij} \mid \mathcal{Y}_k\right] = \frac{\bar{E}\left[\bar{\Lambda}_k J_k^{ij} \mid \mathcal{Y}_k\right]}{\bar{E}\left[\bar{\Lambda}_k \mid \mathcal{Y}_k\right]} := \frac{\sigma(J^{ij})_k}{\langle q_k, \mathbf{1}\rangle}.$$

We wish to know how $\sigma(J^{ij})_k$ is updated as time passes by and new information arrives. However, there does not exist a recursion formula for $\sigma(J^{ij})_k$. Instead, we consider a vector process $\sigma(J^{ij}X)_k := \bar{E}[\bar{\Lambda}_k J_k^{ij} X_k | \mathcal{Y}_k]$ for which recursive formulae can be derived. We then readily obtain the quantity of interest, namely $\sigma(J^{ij})_k$, since $\sigma(J^{ij})_k = \langle \sigma(J^{ij}X)_k, \mathbf{1}\rangle$. We have the following result:

Lemma 1.

$$\sigma(J^{ij}X)_{k+1} = B(Y_{k+1})A\sigma(J^{ij}X)_k + \Big(M \sum_{s=1}^{M} c_{sj} \langle Y_{k+1}, f_s\rangle\Big) \langle q_k, e_i\rangle a_{ji} e_j.$$

Proof. See the Appendix. □

Similarly, we consider the best (mean square) estimates of O_k^i and T_k^{ij} given $\mathcal{Y}_k$:

$$E\left[O_k^i \mid \mathcal{Y}_k\right] = \frac{\bar{E}\left[\bar{\Lambda}_k O_k^i \mid \mathcal{Y}_k\right]}{\bar{E}\left[\bar{\Lambda}_k \mid \mathcal{Y}_k\right]} := \frac{\sigma(O^i)_k}{\langle q_k, \mathbf{1}\rangle},$$

$$E\left[T_k^{ij} \mid \mathcal{Y}_k\right] = \frac{\bar{E}\left[\bar{\Lambda}_k T_k^{ij} \mid \mathcal{Y}_k\right]}{\bar{E}\left[\bar{\Lambda}_k \mid \mathcal{Y}_k\right]} := \frac{\sigma(T^{ij})_k}{\langle q_k, \mathbf{1}\rangle}.$$

Recursive formulae for the processes $\sigma(O^iX)_k := \bar{E}[\bar{\Lambda}_k O_k^i X_k \mid \mathcal{Y}_k]$ and $\sigma(T^{ij}X)_k := \bar{E}[\bar{\Lambda}_k T_k^{ij} X_k | \mathcal{Y}_k]$ are as follows:

Lemma 2.

$$\sigma(O^iX)_{k+1} = B(Y_{k+1})A\sigma(O^iX)_k + \langle q_k, e_i\rangle B(Y_{k+1})Ae_i,$$

$$\sigma(T^{ij}X)_{k+1} = B(Y_{k+1})A\sigma(T^{ij}X)_k + Mc_{ji}\langle Y_{k+1}, f_j\rangle \langle Aq_k, e_i\rangle e_i.$$

Proof. See the Appendix. □

Note that $\sigma(O^i)_k = \langle \sigma(O^i X)_k, \mathbf{1}\rangle$ and $\sigma(T^{ij})_k = \langle \sigma(T^{ij} X)_k, \mathbf{1}\rangle$.

Remark 3. Define $O1^i_k := \sum_{j=1}^M T^{ij}_k = O^i_{k+1}$. Then,

$$\begin{aligned}\sigma(O1^i X)_{k+1} &= \sigma(O^i X)_{k+1} + \Big(M \sum_{s=1}^{M} c_{si}\langle Y_k, f_s\rangle\Big)\langle A q_{k-1}, e_i\rangle e_i \\ &= B(Y_{k+1}) A \sigma(O^i X)_k + \langle A q_k, e_i\rangle B(Y_{k+1}) A e_i \\ &\quad + \Big(M \sum_{s=1}^{M} c_{si}\langle Y_k, f_s\rangle\Big)\langle A q_{k-1}, e_i\rangle e_i\end{aligned}$$

and

$$\sigma(O1^i)_k = \sigma(O^i)_k + \Big(M \sum_{s=1}^{M} c_{si}\langle Y_{k+1}, f_s\rangle\Big)\langle A q_k, e_i\rangle.$$

Proof. See the Appendix. □

Our model is determined by parameters We want to determine a new set of parameters $\theta = \{a_{ji}, 1 \le i, j \le N; c_{ji}, 1 \le i \le N, 1 \le j \le M\}, a_{ji} \ge 0, \sum_{j=1}^N a_{ji} = 1, c_{ji} \ge 0, \sum_{j=1}^M c_{ji} = 1\}$. We want to determine a new set of parameters $\hat\theta = \{\hat a_{ji}, 1 \le i, j \le N; \hat c_{ji}, 1 \le i \le N, 1 \le j \le M\}$ given the arrival of new information, which requires maximum likelihood estimation. We proceed by using the so-called EM (Expectation Maximization) algorithm.

Suppose $\{P_\theta, \theta \in \Theta\}$ is a family of probability measures on a measurable space $(\Omega, \mathcal{F})$. Suppose also that there is another σ-field $\mathcal{Y} \subset \mathcal{F}$. The likelihood function for computing an estimate of θ based on information given in $\mathcal{Y}$ is

$$L(\theta) = E_0\Big[\log \frac{dP_\theta}{dP_0} \,\Big|\, \mathcal{Y}\Big].$$

The maximum likelihood estimate (MLE) of θ is then

$$\hat\theta \in \arg\max_{\theta\in\Theta} L(\theta).$$

However, MLE is hard to compute. The expectation maximization (EM) algorithm provides an alternative approximate method

Step 1: Set $p = 0$ and choose $\hat\theta_0$.
Step 2: (E-step) Set $\theta^* = \hat\theta_p$ and compute

$$Q(\theta, \theta^*) = E_{\theta^*}\Big[\log \frac{dP_\theta}{dP_{\theta^*}} \,\Big|\, \mathcal{Y}\Big].$$

Step 3: (M-step) Find

$$\hat{\theta}_{p+1} \in \arg\max_{\theta \in \Theta} Q(\theta, \theta^*).$$

Step 4: Replace p by $p+1$ and repeat from Step 2 until a stopping criterion is satisfied.

As shown in [8], in our case the EM algorithm produces estimates of model parameters as follows. Given the observations up to time k, $\{Y_0, Y_1, \ldots, Y_k\}$, and given the parameter set $\theta = \{a_{ji}, 1 \le i, j \le N; c_{ji}, 1 \le i \le N, 1 \le j \le M\}$, the EM estimates $\hat{a}_{ji}$ are given by

$$\hat{a}_{ji} = \frac{\sigma(J^{ij})_k}{\sigma(O^i)_k}.$$

Similarly, the EM estimates $\hat{c}_{ji}$ are given by

$$\hat{c}_{ji} = \frac{\sigma(T^{ij})_k}{\sigma(O^i)_k + \left(M \sum_{s=1}^{M} c_{si} \langle Y_k, f_s \rangle\right) \langle A q_{k-1}, e_i \rangle}.$$

5.6 Smoothed estimates

Suppose $0 \le k \le T$ and we are given the information $\mathcal{Y}_{0,T} = \sigma\{Y_0, Y_1, \ldots, Y_T\}$. We wish to estimate X_k given $\mathcal{Y}_{0,T}$. From Bayes' Theorem,

$$E\big[X_k \mid \mathcal{Y}_{0,T}\big] = \frac{\bar{E}\big[\bar{\Lambda}_{0,T} X_k \mid \mathcal{Y}_{0,T}\big]}{\bar{E}\big[\bar{\Lambda}_{0,T} \mid \mathcal{Y}_{0,T}\big]},$$

where $\bar{\Lambda}_{0,T} = \prod_{k=0}^{T} \bar{\lambda}_k$, $\bar{\lambda}_k = M \sum_{j=1}^{M} \langle C X_k, f_j \rangle \langle Y_k, f_j \rangle$. As in Remark 1, the denominator is

$$\bar{E}[\bar{\Lambda}_{0,T} \mid \mathcal{Y}_{0,T}] = \langle q_T, \mathbf{1} \rangle,$$

where $q_T = \bar{E}[\bar{\Lambda}_{0,T} X_T \mid \mathcal{Y}_{0,T}]$ and $\mathbf{1} = (1, 1, \ldots, 1)' \in \mathbb{R}^N$. Now,

$$\begin{aligned}
\bar{E}\big[\bar{\Lambda}_{0,T} X_k \mid \mathcal{Y}_{0,T}\big] &= \bar{E}\big[\bar{\Lambda}_{0,k} \bar{\Lambda}_{k+1,T} X_k \mid \mathcal{Y}_{0,T}\big] \\
&= \bar{E}\big[\bar{\Lambda}_{0,k} X_k \bar{E}\big[\bar{\Lambda}_{k+1,T} \mid \mathcal{Y}_{0,T} \vee \mathcal{F}_k\big] \mid \mathcal{Y}_{0,T}\big],
\end{aligned}$$

where $\bar{\Lambda}_{k+1,T} = \prod_{l=k+1}^{T} \bar{\lambda}_l$. Consider $\bar{E}[\bar{\Lambda}_{k+1,T} \mid \mathcal{Y}_{0,T} \vee \mathcal{F}_k]$, which equals $\bar{E}[\bar{\Lambda}_{k+1,T} | \mathcal{Y}_{0,T} \vee X_k]$ using the Markov property. Write $v_k = (v_k^1, \ldots, v_k^N)'$, where $v_k^i := \bar{E}[\bar{\Lambda}_{k+1,T} \mid \mathcal{Y}_{0,T} \vee \{X_k = e_i\}]$.

Lemma 3. *v satisfies the backwards dynamics, dual to q, of the form*

$$v_k = A' B(Y_{k+1}) v_{k+1}.$$

Proof. See the Appendix. □

Lemma 4. $v_T = (1, \ldots, 1)' \in \mathbb{R}^N$.

Proof. See the Appendix. □

Remark 4. Since $v_T = \mathbf{1}$, we have $v_k = A'B(Y_{k+1})A'B(Y_{k+2}) \cdots A'B(Y_T)\mathbf{1}$.

Theorem 1. *The unnormalized smoothed estimate is*

$$\bar{E}[\bar{\Lambda}_{0,T} X_k \mid \mathcal{Y}_{0,T}] = \text{diag}\,(q_k \cdot v_k').$$

Proof. See the Appendix. □

It follows that

$$E[X_k \mid \mathcal{Y}_{0,T}] = \frac{\text{diag}\,(q_k \cdot v_k')}{\langle q_T, \mathbf{1} \rangle}.$$

Hence, to estimate $E[X_k \mid \mathcal{Y}_{0,T}]$ we need only know the dynamics of q and v, which are, respectively:

$$q_k = B(Y_k)AB(Y_{k-1})A \cdots B(Y_0)Aq_0,$$

where q_0 is the initial distribution for X_0, and

$$v_k = A'B(Y_{k+1})A'B(Y_{k+2}) \cdots A'B(Y_T) \cdot \mathbf{1}.$$

Given observations $\mathcal{Y}_{0,T} = \sigma\{Y_0, Y_1, \ldots, Y_T\}$, we are interested in the smoothed estimates of the number of jumps, the occupation time and the time spent. Consider first the smoothed estimate $E[J_k^{ij} X_k \mid \mathcal{Y}_{0,T}]$. Using Bayes' theorem,

$$E\big[J_k^{ij} X_k \mid \mathcal{Y}_{0,T}\big] = \frac{\bar{E}\big[\bar{\Lambda}_{0,T} J_k^{ij} X_k \mid \mathcal{Y}_{0,T}\big]}{\bar{E}\big[\bar{\Lambda}_{0,T} \mid \mathcal{Y}_{0,T}\big]}.$$

The numerator is $\bar{E}[\bar{\Lambda}_{0,k} J_k^{ij} X_k \bar{\Lambda}_{k+1,T} \mid \mathcal{Y}_{0,T}]$. Consider the l-th component:

$$\begin{aligned}
&\bar{E}\big[\bar{\Lambda}_{0,k} J_k^{ij} X_k \bar{\Lambda}_{k+1,T} \langle X_k, e_l \rangle \mid \mathcal{Y}_{0,T}\big] \\
&= \bar{E}\big[\bar{\Lambda}_{0,k} J_k^{ij} X_k \bar{E}\big[\bar{\Lambda}_{k+1,T} \mid \mathcal{Y}_{0,T} \vee \{X_k = e_l\}\big] \langle X_k, e_l \rangle \mid \mathcal{Y}_{0,T}\big] \\
&= \bar{E}\big[\bar{\Lambda}_{0,k} J_k^{ij} X_k v_k^l \langle X_k, e_l \rangle \mid \mathcal{Y}_{0,T}\big] \\
&= \bar{E}\big[\bar{\Lambda}_{0,k} J_k^{ij} X_k \langle X_k, e_l \rangle \mid \mathcal{Y}_{0,T}\big] v_k^l.
\end{aligned}$$

Then,

$$\begin{aligned}
\bar{E}\big[\bar{\Lambda}_{0,T} J_k^{ij} X_k \mid \mathcal{Y}_{0,T}\big] &= \sum_{l=1}^{N} \bar{E}\big[\bar{\Lambda}_{0,k} J_k^{ij} \langle X_k, e_l \rangle e_l \mid \mathcal{Y}_{0,T}\big] v_k^l \\
&= \sum_{l=1}^{N} \bar{E}\big[\bar{\Lambda}_{0,k} J_k^{ij} \langle X_k, e_l \rangle \mid \mathcal{Y}_{0,T}\big] v_k^l e_l \\
&= \sum_{l=1}^{N} \langle \bar{E}\big[\bar{\Lambda}_{0,k} J_k^{ij} X_k \mid \mathcal{Y}_{0,T}\big], e_l \rangle v_k^l e_l.
\end{aligned}$$

Recall $\sigma(J^{ij}X)_k = \bar{E}[\bar{\Lambda}_k J_k^{ij} X_k \mid \mathcal{Y}_k]$. We then have

$$\begin{aligned}\bar{E}\big[\bar{\Lambda}_{0,T} J_k^{ij} X_k \mid \mathcal{Y}_{0,T}\big] &= \sum_{l=1}^{N} \langle \sigma(J^{ij}X)_k, e_l\rangle v_k^l e_l \\ &= \sum_{l=1}^{N} \sigma(J^{ij}X)_k^l v_k^l e_l \\ &= \operatorname{diag}\big(\sigma(J^{ij}X)_k \cdot v_k^l\big).\end{aligned}$$

Therefore, $\mathbf{1}'\operatorname{diag}\big(\sigma(J^{ij}X)_k \cdot v_k^l\big) = \langle \sigma(J^{ij}X)_k, v_k\rangle = \bar{E}[\bar{\Lambda}_{0,T} J_k^{ij} \mid \mathcal{Y}_{0,T}]$ is the unnormalized, smoothed estimate of J_k^{ij} given $\mathcal{Y}_{0,T}$.

Given observations $\mathcal{Y}_{0,T} = \sigma\{Y_0, Y_1, \ldots, Y_T\}$, we are interested in $\sigma(J^{ij})_T$.

Theorem 2.

$$\sigma(J^{ij})_T = a_{ji} \sum_{k=1}^{T} \langle q_{k-1}, e_i\rangle \langle v_k, e_j\rangle \Big(M \sum_{s=1}^{M} c_{sj} \langle Y_k, f_s\rangle \Big).$$

Proof. See the Appendix. □

Corollary 1.

$$\sigma(O^i)_T = \sum_{k=1}^{T} \langle q_{k-1}, e_i\rangle \langle v_{k-1}, e_i\rangle$$

Proof. See the Appendix. □

Remark 5. Again by Bayes' Theorem,

$$E\big[T_k^{ij} X_k \mid \mathcal{Y}_{0,T}\big] = \frac{\bar{E}\big[\bar{\Lambda}_{0,T} T_k^{ij} X_k \mid \mathcal{Y}_{0,T}\big]}{\bar{E}\big[\bar{\Lambda}_{0,T} \mid \mathcal{Y}_{0,T}\big]}.$$

As before, $\mathbf{1}'\operatorname{diag}\big((T^{ij}X)_k \cdot v_k'\big) = \bar{E}[\bar{\Lambda}_{0,T} T_k^{ij} \mid \mathcal{Y}_{0,T}] = \langle \sigma(T^{ij}X)_k, v_k\rangle$.

Theorem 3.

$$\sigma(T^{ij})_T = \sum_{k=1}^{T} M c_{ji} \langle Y_k, f_j\rangle \langle A q_{k-1}, e_i\rangle \langle v_k, e_i\rangle.$$

Proof. See the Appendix. □

Corollary 2.

$$\sigma(O1^i)_T = \sum_{k=1}^{T} \Big(M \sum_{s=1}^{M} c_{si} \langle Y_k, f_s\rangle \Big) \langle v_k, e_i\rangle \langle A q_{k-1}, e_i\rangle.$$

Proof. See the Appendix. □

Write $V_{k+1,T} = A'B(Y_{k+1}) \cdots A'B(Y_T)$ so that

$$v_k = v_{k,T}, \text{ where}$$

$$v_{k,T} = V_{k+1,T} \cdot \mathbf{1}.$$

Note that the methods to update smoothed estimates above have required recalculation of all backward estimates v. Following [4], we now note results that provide for more efficient computations.

Lemma 5. *$v_{k,T+1} = V_{k+1,T+1}\mathbf{1}$, where $V_{k+1,T+1} = V_{k+1,T}A'B(Y_{T+1})$.*

From Theorem 2,

$$\begin{aligned}
\sigma(J^{ij})_T &= a_{ji} \sum_{k=1}^{T} \Big(M \sum_{s=1}^{M} c_{sj} \langle Y_k, f_s \rangle \Big) \langle q_{k-1}, e_i \rangle \langle v_k, e_j \rangle \\
&= a_{ji} \sum_{k=1}^{T} \Big(M \sum_{s=1}^{M} c_{sj} \langle Y_k, f_s \rangle \Big) \langle q_{k-1}, e_i \rangle e_j' v_k \\
&= a_{ji} \sum_{k=1}^{T} \Big(M \sum_{s=1}^{M} c_{sj} \langle Y_k, f_s \rangle \Big) \langle q_{k-1}, e_i \rangle e_j' A'B(Y_{k+1}) \cdots A'B(Y_T)\mathbf{1} \\
&= \Gamma_T' \mathbf{1},
\end{aligned}$$

where $\Gamma_T' = a_{ji} \sum_{k=1}^{T} \Big(M \sum_{s=1}^{M} c_{sj} \langle Y_k, f_s \rangle \Big) \langle q_{k-1}, e_i \rangle e_j' A'B(Y_{k+1}) \cdots A'B(Y_T)$.

Lemma 6.

$$\Gamma_{T+1}' = \Gamma_T' A'B(Y_{T+1}) + a_{ji} \Big(M \sum_{s=1}^{M} c_{sj} \langle Y_{T+1}, f_s \rangle \Big) \langle q_T, e_i \rangle e_j'.$$

Proof. See the Appendix. □

Corollary 3. *$\sigma(O^i)_T = K_T'\mathbf{1}$, where*

$$K_T' = \sum_{k=1}^{T} \langle q_{k-1}, e_i \rangle e_i' A'B(Y_k) \cdots A'B(Y_T).$$

Then,

$$K_{T+1}' = K_T' A'B(Y_{T+1}) + \langle q_T, e_i \rangle e_i'.$$

From Theorem 3 we have

$$\begin{aligned}\sigma(T^{ij})_T &= \sum_{k=1}^{T} Mc_{ji}\langle Y_k, f_j\rangle\langle Aq_{k-1}, e_i\rangle\langle v_k, e_i\rangle \\ &= \sum_{k=1}^{T} Mc_{ji}\langle Y_k, f_j\rangle\langle Aq_{k-1}, e_i\rangle e_i' v_k \\ &= \sum_{k=1}^{T} Mc_{ji}\langle Y_k, f_j\rangle\langle Aq_{k-1}, e_i\rangle e_i' A'B(Y_{k+1})\cdots A'B(Y_T)\mathbf{1} \\ &= H_T'\mathbf{1},\end{aligned}$$

where $H_T' = \sum_{k=1}^{T} Mc_{ji}\langle Y_k, f_j\rangle\langle Aq_{k-1}, e_i\rangle e_i' A'B(Y_{k+1})\cdots A'B(Y_T)$.

Lemma 7.

$$H_{T+1}' = H_T' A'B(Y_{T+1})\mathbf{1} + Mc_{ji}\langle Y_{T+1}, f_j\rangle\langle Aq_T, e_i\rangle e_i'.$$

Proof. See the Appendix. □

Corollary 4. *In particular,* $\sigma(O1^i)_T = \Delta_T'\mathbf{1}$*, where*

$$\Delta_T' = \sum_{k=1}^{T}\langle Aq_{k-1}, e_i\rangle(M\sum_{s=1}^{M} c_{si}\langle Y_k, f_s\rangle)e_i' A'B(Y_{k+1})\cdots A'B(Y_T).$$

Then, $\Delta_{T+1}' = \Delta_T' A'B(Y_{T+1}) + \langle Aq_T, e_i\rangle\big(M\sum_{s=1}^{M}\langle Y_T, f_s\rangle\big)e_i'$.

A Appendix

Proof of Lemma 1

$$\begin{aligned}
\sigma(J^{ij}X)_{k+1} &= \bar{E}\big[\bar{\Lambda}_{k+1}J^{ij}_{k+1}X_{k+1} \mid \mathcal{Y}_{k+1}\big] \\
&= \bar{E}\big[\bar{\Lambda}_k\bar{\lambda}_{k+1}(J^{ij}_k + \langle X_k, e_i\rangle\langle X_{k+1}, e_j\rangle)X_{k+1} \mid \mathcal{Y}_{k+1}\big] \\
&= \bar{E}\big[\bar{\Lambda}_k\bar{\lambda}_{k+1}J^{ij}_k X_{k+1} \mid \mathcal{Y}_{k+1}\big] \\
&\quad + \bar{E}\big[\bar{\Lambda}_k\bar{\lambda}_{k+1}\langle X_k, e_i\rangle\langle X_{k+1}, e_j\rangle X_{k+1} \mid \mathcal{Y}_{k+1}\big].
\end{aligned}$$

Now,

$$\begin{aligned}
&\bar{E}\big[\bar{\Lambda}_k\bar{\lambda}_{k+1}J^{ij}_k X_{k+1} \mid \mathcal{Y}_{k+1}\big] \\
&\quad = \bar{E}\Big[\bar{\Lambda}_k\Big(M\sum_{s=1}^{M}\langle CX_{k+1}, f_s\rangle\langle Y_{k+1}, f_s\rangle\Big)J^{ij}_k X_{k+1} \,\Big|\, \mathcal{Y}_{k+1}\Big] \\
&\quad = \sum_{r=1}^{N}\bar{E}\Big[\bar{\Lambda}_k\Big(M\sum_{s=1}^{M}c_{sr}\langle Y_{k+1}, f_s\rangle\Big)\langle X_{k+1}, e_r\rangle J^{ij}_k e_r \,\Big|\, \mathcal{Y}_{k+1}\Big] \\
&\quad = \sum_{r=1}^{N}\bar{E}\big[\bar{\Lambda}_k\langle X_{k+1}, e_r\rangle J^{ij}_k \mid \mathcal{Y}_{k+1}\big]\Big(M\sum_{s=1}^{M}c_{sr}\langle Y_{k+1}, f_s\rangle\Big)e_r \\
&\quad = \sum_{r=1}^{N}\bar{E}\big[\bar{\Lambda}_k\langle AX_k, e_r\rangle J^{ij}_k \mid \mathcal{Y}_{k+1}\big]\Big(M\sum_{s=1}^{M}c_{sr}\langle Y_{k+1}, f_s\rangle\Big)e_r \\
&\qquad + \sum_{r=1}^{N}\bar{E}\big[\bar{\Lambda}_k\langle V_{k+1}, e_r\rangle J^{ij}_k \mid \mathcal{Y}_{k+1}\big]\Big(M\sum_{s=1}^{M}c_{sr}\langle Y_{k+1}, f_s\rangle\Big)e_r \\
&\quad = \sum_{r=1}^{N}\langle\bar{E}\big[\bar{\Lambda}_k J^{ij}_k AX_k \mid \mathcal{Y}_{k+1}\big], e_r\rangle\Big(M\sum_{s=1}^{M}c_{sr}\langle Y_{k+1}, f_s\rangle\Big)e_r \\
&\quad = \sum_{r=1}^{N}\langle A\bar{E}\big[\bar{\Lambda}_k J^{ij}_k X_k \mid \mathcal{Y}_k\big], e_r\rangle\Big(M\sum_{s=1}^{M}c_{sr}\langle Y_{k+1}, f_s\rangle\Big)e_r \\
&\quad = \sum_{r=1}^{N}\langle A\sigma(J^{ij}X)_k, e_r\rangle\Big(M\sum_{s=1}^{M}c_{sr}\langle Y_{k+1}, f_s\rangle\Big)e_r \\
&\quad = B(Y_{k+1})A\sigma(J^{ij}X)_k.
\end{aligned}$$

Also,

$$
\begin{aligned}
&\bar{E}\big[\bar{\Lambda}_k\bar{\lambda}_{k+1}\langle X_k, e_i\rangle\langle X_{k+1}, e_j\rangle X_{k+1} \mid \mathcal{Y}_{k+1}\big] = \\
&= \bar{E}\Big[\bar{\Lambda}_k\Big(M\sum_{s=1}^{M}\langle CX_{k+1}, f_s\rangle\langle Y_{k+1}, f_s\rangle\Big)\langle X_k, e_i\rangle\langle X_{k+1}, e_j\rangle X_{k+1} \,\Big|\, \mathcal{Y}_{k+1}\Big] \\
&= \bar{E}\Big[\bar{\Lambda}_k\Big(M\sum_{s=1}^{M}c_{sj}\langle Y_{k+1}, f_s\rangle\Big)\langle X_k, e_i\rangle\langle X_{k+1}, e_j\rangle e_j \,\Big|\, \mathcal{Y}_{k+1}\Big] \\
&= \bar{E}\Big[\bar{\Lambda}_k\Big(M\sum_{s=1}^{M}c_{sj}\langle Y_{k+1}, f_s\rangle\Big)\langle X_k, e_i\rangle\langle AX_k, e_j\rangle e_j \,\Big|\, \mathcal{Y}_{k+1}\Big] \\
&\quad + \bar{E}\Big[\bar{\Lambda}_k\Big(M\sum_{s=1}^{M}c_{sj}\langle Y_{k+1}, f_s\rangle\Big)\langle X_k, e_i\rangle\langle V_{k+1}, e_j\rangle e_j \,\Big|\, \mathcal{Y}_{k+1}\Big] \\
&= \Big(M\sum_{s=1}^{M}c_{sj}\langle Y_{k+1}, f_s\rangle\Big)\bar{E}\Big[\bar{\Lambda}_k\langle X_k, e_i\rangle\langle AX_k, e_j\rangle \,\Big|\, \mathcal{Y}_{k+1}\Big]e_j \\
&\quad + \Big(M\sum_{s=1}^{M}c_{sj}\langle Y_{k+1}, f_s\rangle\Big)\bar{E}\big[\bar{\Lambda}_k\langle X_k, e_i\rangle\langle V_{k+1}, e_j\rangle \mid \mathcal{Y}_{k+1}\big]e_j \\
&= \Big(M\sum_{s=1}^{M}c_{sj}\langle Y_{k+1}, f_s\rangle\Big)\bar{E}\big[\bar{\Lambda}_k\langle X_k, e_i\rangle a_{ji} \mid \mathcal{Y}_{k+1}\big]e_j \\
&= \Big(M\sum_{s=1}^{M}c_{sj}\langle Y_{k+1}, f_s\rangle\Big)\langle \bar{E}\big[\bar{\Lambda}_k X_k \mid \mathcal{Y}_{k+1}\big], e_i\rangle a_{ji}e_j \\
&= \Big(M\sum_{s=1}^{M}c_{sj}\langle Y_{k+1}, f_s\rangle\Big)\langle \bar{E}\big[\bar{\Lambda}_k X_k \mid \mathcal{Y}_{k}\big], e_i\rangle a_{ji}e_j \\
&= \Big(M\sum_{s=1}^{M}c_{sj}\langle Y_{k+1}, f_s\rangle\Big)\langle q_k, e_i\rangle a_{ji}e_j .
\end{aligned}
$$

Therefore,

$$
\sigma(J^{ij}X)_{k+1} = B(Y_{k+1})A\sigma(J^{ij}X)_k + \Big(M\sum_{s=1}^{M}c_{sj}\langle Y_{k+1}, f_s\rangle\Big)\langle q_k, e_i\rangle a_{ji}e_j,
$$

as required. □

Proof of Lemma 2

$$
\begin{aligned}
\sigma(O^iX)_{k+1} &= \bar{E}\big[\bar{\Lambda}_{k+1}O^i_{k+1}X_{k+1} \mid \mathcal{Y}_{k+1}\big] \\
&= \bar{E}\big[\bar{\Lambda}_k\bar{\lambda}_{k+1}(O^i_k + \langle X_k, e_i\rangle)X_{k+1} \mid \mathcal{Y}_{k+1}\big] \\
&= \bar{E}\big[\bar{\Lambda}_k\bar{\lambda}_{k+1}O^i_kX_{k+1} \mid \mathcal{Y}_{k+1}\big] + \bar{E}\big[\bar{\Lambda}_k\bar{\lambda}_{k+1}\langle X_k, e_i\rangle X_{k+1} \mid \mathcal{Y}_{k+1}\big].
\end{aligned}
$$

Now,

$$
\begin{aligned}
&\bar{E}\big[\bar{\Lambda}_k \bar{\lambda}_{k+1} O^i_k X_{k+1} \mid \mathcal{Y}_{k+1}\big] \\
&\quad = \bar{E}\Big[\bar{\Lambda}_k \Big(M \sum_{j=1}^{M} \langle C X_{k+1}, f_j\rangle \langle Y_{k+1}, f_j\rangle \Big) O^i_k X_{k+1} \;\Big|\; \mathcal{Y}_{k+1}\Big] \\
&\quad = \sum_{r=1}^{N} \bar{E}\Big[\bar{\Lambda}_k \Big(M \sum_{j=1}^{M} c_{jr} \langle Y_{k+1}, f_j\rangle \Big) \langle X_{k+1}, e_r\rangle O^i_k e_r \;\Big|\; \mathcal{Y}_{k+1}\Big] e_r \\
&\quad = \sum_{r=1}^{N} \Big(M \sum_{j=1}^{M} c_{jr} \langle Y_{k+1}, f_j\rangle \Big) \bar{E}\big[\bar{\Lambda}_k \langle X_{k+1}, e_r\rangle O^i_k \mid \mathcal{Y}_{k+1}\big] e_r \\
&\quad = \sum_{r=1}^{N} \Big(M \sum_{j=1}^{M} c_{jr} \langle Y_{k+1}, f_j\rangle \Big) \langle \bar{E}\big[\bar{\Lambda}_k A X_k O^i_k \mid \mathcal{Y}_{k+1}\big], e_r\rangle e_r \\
&\qquad + \sum_{r=1}^{N} \Big(M \sum_{j=1}^{M} c_{jr} \langle Y_{k+1}, f_j\rangle \Big) \langle \bar{E}\big[\bar{\Lambda}_k O^i_k V_{k+1} \mid \mathcal{Y}_{k+1}\big], e_r\rangle e_r \\
&\quad = \sum_{r=1}^{N} \Big(M \sum_{j=1}^{M} c_{jr} \langle Y_{k+1}, f_j\rangle \Big) \langle A \bar{E}\big[\bar{\Lambda}_k O^i_k X_k \mid \mathcal{Y}_k\big], e_r\rangle e_r \\
&\quad = \sum_{r=1}^{N} \Big(M \sum_{j=1}^{M} c_{jr} \langle Y_{k+1}, f_j\rangle \Big) \langle A \sigma(O^i X)_k, e_r\rangle e_r \\
&\quad = B(Y_{k+1}) A \sigma(O^i X)_k .
\end{aligned}
$$

Also,

$$
\begin{aligned}
&\bar{E}\big[\bar{\Lambda}_k \bar{\lambda}_{k+1}\langle X_k, e_i\rangle X_{k+1} \mid \mathcal{Y}_{k+1}\big] \\
&\quad = \bar{E}\Big[\bar{\Lambda}_k\Big(M\sum_{j=1}^{M}\langle CX_{k+1}, f_j\rangle\langle Y_{k+1}, f_j\rangle\Big)\langle X_k, e_i\rangle X_{k+1} \Bigm| \mathcal{Y}_{k+1}\Big] \\
&\quad = \sum_{r=1}^{N}\bar{E}\Big[\bar{\Lambda}_k\Big(M\sum_{j=1}^{M} c_{jr}\langle Y_{k+1}, f_j\rangle\Big)\langle X_{k+1}, e_r\rangle\langle X_k, e_i\rangle e_r \Bigm| \mathcal{Y}_{k+1}\Big] \\
&\quad = \sum_{r=1}^{N}\Big(M\sum_{j=1}^{M} c_{jr}\langle Y_{k+1}, f_j\rangle\Big)\bar{E}\big[\bar{\Lambda}_k\langle X_{k+1}, e_r\rangle\langle X_k, e_i\rangle \mid \mathcal{Y}_{k+1}\big]e_r \\
&\quad = \sum_{r=1}^{N}\Big(M\sum_{j=1}^{M} c_{jr}\langle Y_{k+1}, f_j\rangle\Big)\bar{E}\big[\bar{\Lambda}_k\langle AX_k, e_r\rangle\langle X_k, e_i\rangle \mid \mathcal{Y}_{k+1}\big]e_r \\
&\qquad + \sum_{r=1}^{N}\Big(M\sum_{j=1}^{M} c_{jr}\langle Y_{k+1}, f_j\rangle\Big)\bar{E}\big[\bar{\Lambda}_k\langle V_{k+1}, e_r\rangle\langle X_k, e_i\rangle \mid \mathcal{Y}_{k+1}\big]e_r \\
&\quad = \sum_{r=1}^{N}\Big(M\sum_{j=1}^{M} c_{jr}\langle Y_{k+1}, f_j\rangle\Big)\bar{E}\big[\bar{\Lambda}_k a_{ri}\langle X_k, e_i\rangle \mid \mathcal{Y}_{k+1}\big]e_r \\
&\quad = \sum_{r=1}^{N}\Big(M\sum_{j=1}^{M} c_{jr}\langle Y_{k+1}, f_j\rangle\Big)\bar{E}\big[\bar{\Lambda}_k\langle X_k, e_i\rangle \mid \mathcal{Y}_{k+1}\big]a_{ri}e_r \\
&\quad = \sum_{r=1}^{N}\Big(M\sum_{j=1}^{M} c_{jr}\langle Y_{k+1}, f_j\rangle\Big)\langle q_k, e_i\rangle a_{ri}e_r \\
&\quad = \langle q_k, e_i\rangle B(Y_{k+1})Ae_i.
\end{aligned}
$$

We follow the same procedure to obtain the recursion for the dynamics of the vector process $\sigma(T^{ij}X)_k$. □

Proof of Remark 5.2.

$$
\begin{aligned}
\sigma(O1^i X)_{k+1} - \bar{E}\big[\bar{\Lambda}_{k+1}O1^i_{k+1}X_{k+1} \mid \mathcal{Y}_{k+1}\big] &= \bar{E}\big[\bar{\Lambda}_{k+1}O^i_{k+2}X_{k+1} \mid \mathcal{Y}_{k+1}\big] \\
&= \bar{E}\big[\bar{\Lambda}_{k+1}(O^i_{k+1} + \langle X_{k+1}, e_i\rangle)X_{k+1} \mid \mathcal{Y}_{k+1}\big] \\
&= \bar{E}\big[\bar{\Lambda}_{k+1}O^i_{k+1}X_{k+1}|\mathcal{Y}_{k+1}\big] + \bar{E}\big[\bar{\Lambda}_{k+1}\langle X_{k+1}, e_i\rangle X_{k+1} \mid \mathcal{Y}_{k+1}\big] \\
&= \sigma(O^i X)_{k+1} + \bar{E}\big[\bar{\Lambda}_{k+1}\langle X_{k+1}, e_i\rangle X_{k+1} \mid \mathcal{Y}_{k+1}\big].
\end{aligned}
$$

Now,

$$
\begin{aligned}
&\bar{E}\big[\bar{\Lambda}_{k+1}\langle X_{k+1}, e_i\rangle X_{k+1} \mid \mathcal{Y}_{k+1}\big] = \\
&\quad = \bar{E}\Big[\bar{\Lambda}_k\Big(M\sum_{s=1}^{M}\langle CX_{k+1}, f_s\rangle\langle Y_{k+1}, f_s\rangle\Big)\langle X_{k+1}, e_i\rangle X_{k+1} \;\Big|\; \mathcal{Y}_{k+1}\Big] \\
&\quad = \bar{E}\Big[\bar{\Lambda}_k\Big(M\sum_{s=1}^{M}\langle CX_{k+1}, f_s\rangle\langle Y_{k+1}, f_s\rangle\Big)\langle X_{k+1}, e_i\rangle e_i \;\Big|\; \mathcal{Y}_{k+1}\Big] \\
&\quad = \bar{E}\Big[\bar{\Lambda}_k\Big(M\sum_{s=1}^{M}\langle CX_{k+1}, f_s\rangle\langle Y_{k+1}, f_s\rangle\Big)\langle AX_k + V_{k+1}, e_i\rangle e_i \;\Big|\; \mathcal{Y}_{k+1}\Big] \\
&\quad = \bar{E}\Big[\bar{\Lambda}_k\Big(M\sum_{s=1}^{M}\langle CX_{k+1}, f_s\rangle\langle Y_{k+1}, f_s\rangle\Big)\langle AX_k, e_i\rangle e_i \;\Big|\; \mathcal{Y}_{k+1}\Big] \\
&\qquad + \bar{E}\Big[\bar{\Lambda}_k\Big(M\sum_{s=1}^{M}\langle CX_{k+1}, f_s\rangle\langle Y_{k+1}, f_s\rangle\Big)\langle V_{k+1}, e_i\rangle e_i \;\Big|\; \mathcal{Y}_{k+1}\Big] \\
&\quad = \Big(M\sum_{s=1}^{M}\langle CX_{k+1}, f_s\rangle\langle Y_{k+1}, f_s\rangle\Big)\bar{E}\big[\bar{\Lambda}_k\langle AX_k, e_i\rangle e_i \mid \mathcal{Y}_{k+1}\big] \\
&\quad = \Big(M\sum_{s=1}^{M}\langle CX_{k+1}, f_s\rangle\langle Y_{k+1}, f_s\rangle\Big)\langle A\bar{E}\big[\bar{\Lambda}_k X_k \mid \mathcal{Y}_{k+1}\big], e_i\rangle e_i \\
&\quad = \Big(M\sum_{s=1}^{M}\langle CX_{k+1}, f_s\rangle\langle Y_{k+1}, f_s\rangle\Big)\langle Aq_k, e_i\rangle e_i.
\end{aligned}
$$

The result follows. □

Proof of Lemma 3

$$
\begin{aligned}
v_k^i &= \bar{E}\big[\bar{\Lambda}_{k+1,T} \mid \mathcal{Y}_{0,T} \vee \{X_k = e_i\}\big] \\
&= \bar{E}\big[\bar{\Lambda}_{k+2,T}\bar{\lambda}_{k+1} \mid \mathcal{Y}_{0,T} \vee \{X_k = e_i\}\big] \\
&= \bar{E}\Big[\bar{\Lambda}_{k+2,T}\Big(M\sum_{j=1}^{M}\langle CX_{k+1}, f_j\rangle\langle Y_{k+1}, f_j\rangle\Big) \;\Big|\; \mathcal{Y}_{0,T} \vee \{X_k = e_i\}\Big] \\
&= \sum_{l=1}^{N}\bar{E}\Big[\bar{\Lambda}_{k+2,T}\Big(M\sum_{j=1}^{M}c_{jl}\langle Y_{k+1}, f_j\rangle)\langle X_{k+1}, e_l\rangle \;\Big|\; \mathcal{Y}_{0,T} \vee \{X_k = e_i\}\Big] \\
&= \sum_{l=1}^{N}\bar{E}\big[\bar{\Lambda}_{k+2,T}\langle X_{k+1}, e_l\rangle \mid \mathcal{Y}_{0,T} \vee \{X_k = e_i\}\big]\Big(M\sum_{j=1}^{M}c_{jl}\langle Y_{k+1}, f_j\rangle\Big)
\end{aligned}
$$

$$
\begin{aligned}
&= \sum_{l=1}^{M} \bar{E}\big[\langle X_{k+1}, e_l\rangle \\
&\qquad \times \bar{E}\big[\bar{\Lambda}_{k+2,T} \mid \mathcal{Y}_{0,T} \vee \{X_k = e_i\} \vee \{X_k = e_l\}\big] \mid \mathcal{Y}_{0,T} \vee \{X_k = e_i\}\big] \\
&\quad \times \Big(M \sum_{j=1}^{M} c_{jl}\langle Y_{k+1}, f_j\rangle\Big) \\
&= \sum_{l=1}^{N} \bar{E}\big[\langle X_{k+1}, e_l\rangle v_{k+1}^{l} \mid \mathcal{Y}_{0,T} \vee \{X_k = e_i\}\big]\Big(M \sum_{j=1}^{M} c_{jl}\langle Y_{k+1}, f_j\rangle\Big) \\
&= \sum_{l=1}^{N} \bar{E}\big[\langle X_{k+1}, e_l\rangle \mid \mathcal{Y}_{0,T} \vee \{X_k = e_i\}\big] v_{k+1}^{l}\Big(M \sum_{j=1}^{M} c_{jl}\langle Y_{k+1}, f_j\rangle\Big) \\
&= \sum_{l=1}^{N} \bar{E}\big[\langle AX_k + V_{k+1}, e_l\rangle \mid \mathcal{Y}_{0,T} \vee \{X_k = e_i\}\big] v_{k+1}^{l}\Big(M \sum_{j=1}^{M} c_{jl}\langle Y_{k+1}, f_j\rangle\Big) \\
&= \sum_{l=1}^{N} \bar{E}\big[\langle AX_k, e_l\rangle \mid \mathcal{Y}_{0,T} \vee \{X_k = e_i\}\big] v_{k+1}^{l}\Big(M \sum_{j=1}^{M} c_{jl}\langle Y_{k+1}, f_j\rangle\Big) \\
&\quad + \sum_{l=1}^{N} \bar{E}\big[\langle V_{k+1}, e_l\rangle \mid \mathcal{Y}_{0,T} \vee \{X_k = e_i\}\big] v_{k+1}^{l}\Big(M \sum_{j=1}^{M} c_{jl}\langle Y_{k+1}, f_j\rangle\Big) \\
&= \sum_{l=1}^{N} \bar{P}\big(X_{k+1} = e_l \mid \mathcal{Y}_{0,T} \vee \{X_k = e_i\}\big) v_{k+1}^{l}\Big(M \sum_{j=1}^{M} c_{jl}\langle Y_{k+1}, f_j\rangle\Big) \\
&\quad + \sum_{l=1}^{N} \langle \bar{E}\big[V_{k+1} \mid \mathcal{Y}_{0,T} \vee \{X_k = e_i\}\big], e_l\rangle v_{k+1}^{l}\Big(M \sum_{j=1}^{M} c_{jl}\langle Y_{k+1}, f_j\rangle\Big) \\
&= \sum_{l=1}^{N} \bar{P}\big(X_{k+1} = e_l \mid X_k = e_i\big) v_{k+1}^{l}\Big(M \sum_{j=1}^{M} c_{jl}\langle Y_{k+1}, f_j\rangle\Big) \\
&\quad + \sum_{l=1}^{N} \langle \bar{E}\big[V_{k+1} \mid X_k = e_i\big], e_l\rangle v_{k+1}^{l}\Big(M \sum_{j=1}^{M} c_{jl}\langle Y_{k+1}, f_j\rangle\Big) \\
&= \sum_{l=1}^{N} a_{li} v_{k+1}^{l}\Big(M \sum_{j=1}^{M} c_{jl}\langle Y_{k+1}, f_j\rangle\Big) \\
&\quad + \sum_{l=1}^{N} \langle \bar{E}\big[\bar{E}[V_{k+1} \mid \mathcal{F}_k] \mid X_k = e_i\big], e_l\rangle v_{k+1}^{l}\Big(M \sum_{j=1}^{M} c_{jl}\langle Y_{k+1}, f_j\rangle\Big) \\
&= \sum_{l=1}^{N} a_{li} v_{k+1}^{l}\Big(M \sum_{j=1}^{M} c_{jl}\langle Y_{k+1}, f_j\rangle\Big).
\end{aligned}
$$

It follows that $v_k = A'B(Y_{k+1})v_{k+1}$, as required. $\square$

Proof of Lemma 4

Consider the j-th component of v_{T-1}:

$$\begin{aligned}
v^j_{T-1} &= \bar{E}\big[\bar{\Lambda}_{T,T} \mid \mathcal{Y}_{0,T} \vee \{X_{T-1} = e_j\}\big] = \bar{E}\big[\bar{\lambda}_T \mid \mathcal{Y}_{0,T} \vee \{X_{T-1} = e_j\}\big] \\
&= \bar{E}\Big[M \sum_{l=1}^{M} \langle CX_T, f_l\rangle \langle Y_T, f_l\rangle \,\Big|\, \mathcal{Y}_{0,T} \vee \{X_{T-1} = e_j\}\Big] \\
&= M \sum_{j=1}^{M} \bar{E}\big[\langle CX_T, f_l\rangle \mid \mathcal{Y}_{0,T} \vee \{X_{T-1} = e_j\}\big]\langle Y_T, f_l\rangle \\
&= M \sum_{j=1}^{M} \bar{E}\Big[\sum_{i=1}^{N} c_{li}\langle X_T, e_i\rangle \,\Big|\, \mathcal{Y}_{0,T} \vee \{X_{T-1} = e_j\}\Big]\langle Y_T, f_l\rangle \\
&= \sum_{i=1}^{N} \bar{E}\big[\langle X_T, e_i\rangle \mid \mathcal{Y}_{0,T} \vee \{X_{T-1} = e_j\}\big]\Big(M \sum_{l=1}^{M} c_{li}\langle Y_T, f_l\rangle\Big) \\
&= \sum_{i=1}^{N} \bar{E}\big[\langle X_T, e_i\rangle \mid \{X_{T-1} = e_j\}\big]\Big(M \sum_{l=1}^{M} c_{li}\langle Y_T, f_l\rangle\Big) \\
&= \sum_{i=1}^{N} \bar{P}\big(X_T = e_i \mid X_{T-1} = e_j\big)\Big(M \sum_{l=1}^{M} c_{li}\langle Y_T, f_l\rangle\Big) \\
&= \sum_{i=1}^{N} a_{ij}\Big(M \sum_{l=1}^{M} c_{li}\langle Y_T, f_l\rangle\Big).
\end{aligned}$$

It follows that $v_{T-1} = A'B(Y_T)\mathbf{1}$. $\square$

Proof of Theorem 1

$$\begin{aligned}
\bar{E}\big[\bar{\Lambda}_{0,T}X_k \mid \mathcal{Y}_{0,T}\big] &= \sum_{i=1}^{N} \bar{E}\big[\bar{\Lambda}_{0,T}\langle X_k, e_i\rangle X_k \mid \mathcal{Y}_{0,T}\big] \\
&= \sum_{i=1}^{N} \bar{E}\big[\bar{\Lambda}_{0,T}\langle X_k, e_i\rangle \mid \mathcal{Y}_{0,T}\big]e_i.
\end{aligned}$$

Consider the i-th component:

$$\begin{aligned}
\bar{E}\big[\bar{\Lambda}_{0,T}\langle X_k, e_i\rangle \mid \mathcal{Y}_{0,T}\big] &= \bar{E}\big[\bar{\Lambda}_{0,k}\bar{\Lambda}_{k+1,T}\langle X_k, e_i\rangle \mid \mathcal{Y}_{0,T}\big] \\
&= \bar{E}\big[\bar{\Lambda}_{0,k}\bar{E}\big[\bar{\Lambda}_{k+1,T} \mid \mathcal{Y}_{0,T} \vee \{X_k = e_i\}\big]\langle X_k, e_i\rangle \mid \mathcal{Y}_{0,T}\big] \\
&= \bar{E}\big[\bar{\Lambda}_{0,k}v^i_k\langle X_k, e_i\rangle \mid \mathcal{Y}_{0,T}\big] \\
&= \bar{E}\big[\bar{\Lambda}_{0,k}\langle X_k, e_i\rangle \mid \mathcal{Y}_{0,T}\big]v^i_k \\
&= \langle \bar{E}\big[\bar{\Lambda}_{0,k}X_k \mid \mathcal{Y}_{0,T}\big], e_i\rangle v^i_k \\
&= q^i_k v^i_k,
\end{aligned}$$

where $q_k^i := \langle \bar{E}[\bar{\Lambda}_k X_k \mid \mathcal{Y}_k], e_i \rangle$.

Therefore, $\bar{E}[\bar{\Lambda}_{0,T} X_k \mid \mathcal{Y}_{0,T}] = \sum_{i=1}^N q_k^i v_k^i e_i = \operatorname{diag}(q_k \cdot v_k')$. □

Proof of Theorem 2

$$
\begin{aligned}
&\langle \sigma(J^{ij}X)_{k+1}, v_{k+1} \rangle = \\
&= \langle B(Y_{k+1}) A \sigma(J^{ij}X)_k + \Big(M \sum_{s=1}^M c_{sj} \langle Y_{k+1}, f_s \rangle \Big) \langle q_k, e_i \rangle a_{ji} e_j, v_{k+1} \rangle \\
&= \langle B(Y_{k+1}) A \sigma(J^{ij}X)_k, v_{k+1} \rangle + \Big\langle \Big(M \sum_{s=1}^M c_{sj} \langle Y_{k+1}, f_s \rangle \Big) \langle q_k, e_i \rangle a_{ji} e_j, v_{k+1} \Big\rangle \\
&= \langle B(Y_{k+1}) A \sigma(J^{ij}X)_k, v_{k+1} \rangle + \Big(M \sum_{s=1}^M c_{sj} \langle Y_{k+1}, f_s \rangle \Big) \langle q_k, e_i \rangle a_{ji} \langle v_{k+1}, e_j \rangle \\
&= \langle \sigma(J^{ij}X)_k, A' B(Y_{k+1}) v_{k+1} \rangle + \Big(M \sum_{s=1}^M c_{sj} \langle Y_{k+1}, f_s \rangle \Big) \langle q_k, e_i \rangle \langle v_{k+1}, e_j \rangle a_{ji} \\
&= \langle \sigma(J^{ij}X)_k, v_k \rangle + \Big(M \sum_{s=1}^M c_{sj} \langle Y_{k+1}, f_s \rangle \Big) \langle q_k, e_i \rangle \langle v_{k+1}, e_j \rangle a_{ji}.
\end{aligned}
$$

That is,

$$
\begin{aligned}
&\langle \sigma(J^{ij}X)_{k+1}, v_{k+1} \rangle - \langle \sigma(J^{ij}X)_k, v_k \rangle \\
&\quad = \Big(M \sum_{s=1}^M c_{sj} \langle Y_{k+1}, f_s \rangle \Big) \langle q_k, e_i \rangle \langle v_{k+1}, e_j \rangle a_{ji}.
\end{aligned}
$$

Since $J_0^{ij} = 0$ and $v_T = \mathbf{1}$,

$$
\begin{aligned}
&\sum_{k=0}^{T-1} \big[\langle \sigma(J^{ij}X)_{k+1}, v_{k+1} \rangle - \langle \sigma(J^{ij}X)_k, v_k \rangle \big] = \langle \sigma(J^{ij}X)_T, v_T \rangle - \langle \sigma(J^{ij}X)_0, v_0 \rangle \\
&= \langle \sigma(J^{ij}X)_T, v_T \rangle = \langle \sigma(J^{ij}X)_T, \mathbf{1} \rangle = \sigma(J^{ij})_T.
\end{aligned}
$$

Hence,

$$
\begin{aligned}
\sigma(J^{ij})_T &= \sum_{k=0}^{T-1} a_{ji} \Big(M \sum_{s=1}^M c_{sj} \langle Y_{k+1}, f_s \rangle \Big) \langle q_k, e_i \rangle \langle v_{k+1}, e_j \rangle \\
&= \sum_{k=1}^{T} a_{ji} \Big(M \sum_{s=1}^M c_{sj} \langle Y_k, f_s \rangle \Big) \langle q_{k-1}, e_i \rangle \langle v_k, e_j \rangle. \quad \square
\end{aligned}
$$

Proof of Corollary 1

Since $\sigma(O^i)_T = \sum_{j=1}^N \sigma(J^{ij})_T$, we have

$$\begin{aligned}\sigma(O^i)_T &= \sum_{j=1}^{N} \sigma(J^{ij})_T = \sum_{j=1}^{N} a_{ji} \sum_{k=1}^{T} \langle q_{k-1}, e_i\rangle \langle v_k, e_j\rangle \Big(M \sum_{s=1}^{M} c_{sj} \langle Y_k, f_s\rangle \Big) \\ &= \sum_{k=1}^{T} \langle q_{k-1}, e_i\rangle \sum_{j=1}^{N} a_{ji} \langle v_k, e_j\rangle \Big(M \sum_{s=1}^{M} c_{sj} \langle Y_k, f_s\rangle \Big) \\ &= \sum_{k=1}^{T} \langle q_{k-1}, e_i\rangle \sum_{j=1}^{N} v_k^j a_{ji} \Big(M \sum_{s=1}^{M} c_{sj} \langle Y_k, f_s\rangle \Big) \\ &= \sum_{k=1}^{T} \langle q_{k-1}, e_i\rangle \langle A' B(Y_{k+1}) v_k, e_i\rangle \\ &= \sum_{k=1}^{T} \langle q_{k-1}, e_i\rangle \langle v_{k-1}, e_i\rangle. \quad \square\end{aligned}$$

Proof of Theorem 3

$$\begin{aligned}&\langle \sigma(T^{ij}X)_{k+1}, v_{k+1}\rangle = \\ &= \langle B(Y_{k+1}) A \sigma(T^{ij}X)_k + M c_{ji} \langle Y_{k+1}, f_j\rangle \langle A q_k, e_i\rangle e_i, v_{k+1}\rangle \\ &\text{(by Theorem 2.3)} \\ &= \langle B(Y_{k+1}) A \sigma(T^{ij}X)_k, v_{k+1}\rangle + \langle M c_{ji} \langle Y_{k+1}, f_j\rangle \langle A q_k, e_i\rangle e_i, v_{k+1}\rangle \\ &= \langle B(Y_{k+1}) A \sigma(T^{ij}X)_k, v_{k+1}\rangle + M c_{ji} \langle Y_{k+1}, f_j\rangle \langle A q_k, e_i\rangle \langle v_{k+1}, e_i\rangle \\ &= \langle \sigma(T^{ij}X)_k, A' B(Y_{k+1}) v_{k+1}\rangle + M c_{ji} \langle Y_{k+1}, f_j\rangle \langle A q_k, e_i\rangle \langle v_{k+1}, e_i\rangle \\ &= \langle \sigma(T^{ij}X)_k, v_k\rangle + M c_{ji} \langle Y_{k+1}, f_j\rangle \langle A q_k, e_i\rangle \langle v_{k+1}, e_i\rangle.\end{aligned}$$

That is,

$$\langle \sigma(T^{ij}X)_{k+1}, v_{k+1}\rangle - \langle \sigma(T^{ij}X)_k, v_k\rangle = M c_{ji} \langle Y_{k+1}, f_j\rangle \langle A q_k, e_i\rangle \langle v_{k+1}, e_j\rangle.$$

Since $T_0^{ij} = 0$ and $v_T = 1$,

$$\begin{aligned}&\sum_{k=0}^{T-1} [\langle \sigma(T^{ij}X)_{k+1}, v_{k+1}\rangle - \langle \sigma(T^{ij}X)_k, v_k\rangle] = \langle \sigma(T^{ij}X)_T, v_T\rangle - \langle \sigma(T^{ij}X)_0, v_0\rangle \\ &= \langle \sigma(T^{ij}X)_T, v_T\rangle = \langle \sigma(T^{ij}X)_T, 1\rangle = \sigma(T^{ij})_T.\end{aligned}$$

Hence,

$$\begin{aligned}\sigma(T^{ij})_T &= \sum_{k=0}^{T-1} M c_{ji} \langle Y_{k+1}, f_j\rangle \langle A q_k, e_i\rangle \langle v_{k+1}, e_i\rangle \\ &= \sum_{k=1}^{T} M c_{ji} \langle Y_k, f_j\rangle \langle A q_{k-1}, e_i\rangle \langle v_k, e_i\rangle. \quad \square\end{aligned}$$

Proof of Corollary 2
Since

$$\sigma(O1^i)_T = \sum_{j=1}^{M} \sigma(T^{ij})_T,$$

we have

$$\begin{aligned}
\sigma(O1^i)_T &= \sum_{j=1}^{M} \sigma(T^{ij})_T = \sum_{j=1}^{M}\sum_{k=1}^{T} M c_{ji}\langle Y_k, f_j\rangle\langle Aq_{k-1}, e_i\rangle\langle v_k, e_i\rangle \\
&= \sum_{k=1}^{T}\langle v_k, e_i\rangle\langle Aq_{k-1}, e_i\rangle\Big(M\sum_{j=1}^{M} c_{ji}\langle Y_k, f_j\rangle\Big). \quad \square
\end{aligned}$$

Proof of Lemma 6

$$\begin{aligned}
\sigma(J^{ij})_{T+1} &= a_{ji}\sum_{k=1}^{T+1}\Big(M\sum_{s=1}^{M} c_{sj}\langle Y_k, f_s\rangle\Big)\langle q_{k-1}, e_i\rangle e_j' \\
&\quad \times A'B(Y_{k+1})\cdots A'B(Y_T)A'B(Y_{T+1})\mathbf{1} \\
&= a_{ji}\sum_{k=1}^{T}\Big(M\sum_{s=1}^{M} c_{sj}\langle Y_k, f_s\rangle\Big)\langle q_{k-1}, e_i\rangle e_j' A'B(Y_{k+1})\cdots A'B(Y_T)\mathbf{1} \\
&\quad + a_{ji}\Big(M\sum_{s=1}^{M} c_{sj}\langle Y_{T+1}, f_s\rangle\Big)\langle q_T, e_i\rangle e_j'\mathbf{1} \\
&= \Gamma_T' A'B(Y_{T+1}) + a_{ji}\Big(M\sum_{s=1}^{M} c_{sj}\langle Y_{T+1}, f_s\rangle\Big)\langle q_T, e_i\rangle e_j' \\
&= \Gamma_{T+1}'\mathbf{1}.
\end{aligned}$$

The result follows. $\square$

Proof of Lemma 7

$$\begin{aligned}
\sigma(T^{ij})_{T+1} &= \sum_{k=1}^{T+1} M c_{ji}\langle Y_k, f_j\rangle\langle Aq_{k-1}, e_i\rangle e_i' A'B(Y_{k+1})\cdots A'B(Y_T)A'B(Y_{T+1})\mathbf{1} \\
&= \sum_{k=1}^{T} M c_{ji}\langle Y_k, f_j\rangle\langle Aq_{k-1}, e_i\rangle e_i' A'B(Y_{k+1})\cdots A'B(Y_T)A'B(Y_{T+1})\mathbf{1} \\
&\quad + M c_{ji}\langle Y_{T+1}, f_j\rangle\langle Aq_T, e_i\rangle e_i'\mathbf{1} \\
&= H_T' A'B(Y_{T+1})\mathbf{1} + M c_{ji}\langle Y_{T+1}, f_j\rangle\langle Aq_T, e_i\rangle e_i'\mathbf{1} \\
&= H_{T+1}'\mathbf{1}.
\end{aligned}$$

The result follows. $\square$

References

1. Carty, L.V. and J.S. Fons (1994). "Measuring changes in corporate credit quality". *Journal of Fixed Income*, 4(1): 27-41.
2. Carty, L.V. and D. Lieberman (1997). "Historical default rates of corporate bond issuers, 1920-1996". *Moody's Investors Service.*
3. Elliott, R.J. (1994). "Exact adaptive filters for Markov chains observed in Gaussian noise". *Automatica*, 30(9): 1399-1408.
4. Elliott, R.J. (1997). "Improved smoothers for discrete time HMM parameter estimation". University of Alberta and University of Adelaide Working Paper, 1997.
5. Elliott, R.J., Aggoun, L. and J.B. Moore (1995). *Hidden Markov Models. Estimation and Control.* Springer-Verlag.
6. Jarrow, R.A. and S.M. Turnbull (1995). "Pricing derivatives on financial securities subject to credit risk". *The Journal of Finance*, L(1): 53-85.
7. Jarrow, R.A., Lando, D. and S.M. Turnbull (1997). "A Markov model for the term structure of credit risk spreads". *The Review of Financial Studies*, 10(2): 481-523.
8. Korolkiewicz, M.W. (2004). *Applications of Hidden Markov Chains to Credit Risk Modelling.* PhD Thesis, University of Alberta.
9. Lando, D. and T. Skodeberg (2002). "Analyzing rating transitions and rating drift with continuous observations". *The Journal of Banking and Finance*, 26: 423-444.

6

Expected Shortfall Under a Model With Market and Credit Risks

Kin Bong Siu and Hailiang Yang

Department of Statistics and Actuarial Science
The University of Hong Kong
Pokfulam Road, Hong Kong
h0010297@hkusua.hku.hk (K.B. Siu)
hlyang@hkusua.hku.hk (H. Yang)

Summary. Value-at-Risk (VaR), due to its simplicity and ease of interpretability, has become a popular risk measure in finance nowadays. However, recent research find that VaR is not a coherent risk measure and cannot incorporate the loss beyond VaR or tail risk. This chapter considers expected shortfall (ES) as an alternative risk measure. We consider a portfolio subject to both market and credit risks. We model the credit rating using a Markov chain. Thus our model can be treated as a Markovian regime-switching model. We also propose a weak Markov chain model which can take into account the dependency of the risks. Expressions for VaR, ES and numerical results are presented to illustrate the proposed ideas.

Key words: Value at Risk, expected shortfall, market risk, credit risk, credit ranking, Markov chain, weak Markov chain, coherent risk measure.

6.1 Introduction

It has been an aim for a long time in finance to have an appropriate measure for the risk of an investment portfolio. VaR, being simple to interpret, has become more popular in risk management subjects. VaR is generally defined as the possible maximum loss over a given holding period within a pre-defined confidence level (Yamai and Yoshiba, [11]). Artzner et al. [4] defines VaR at 1-α confidence level mathematically as the lower 100 α percentile of the portfolio return distribution:

$$VaR_\alpha(X) = -\inf\{x|P[X \leq x] > \alpha\}$$

Risk managers and regulators have put a lot of efforts on VaR in the early years because of the promise it holds for improving risk management. International bank regulators have also agreed to allow local banks to adopt VaR models to calculate regulatory capital. Although VaR provides fund managers a quick and readily accessible value on their portfolio risk, critics on VaR grew with its popular uses. The main criticism include:

a) VaR is not a coherent risk measure under certain situation, owing to its non sub-additivity.
b) VaR ignores the tail risk, since it disregards the tail distribution beyond its value.
c) The use of VaR allows construction of proxies portfolios having low VaR which resulted from a trade-off of heavy tail loss.
d) Information given by VaR may misled rational investors who wish to maximize expected utility. In particular, employing VaR as the only risk measure is more likely to construct perverse position that would result a larger loss beyond VaR level.

The concept of coherence for a risk measure is introduced by Artzner et al. [3]. They present four desirable properties of risk measures and regarded those risk measures satisfying all four properties as coherent. These four properties are (i) Translation Invariance (ii) Sub-Additivity (iii) Positive Homogeneity and (iv) Monotonicity. All of them have their own practical interpretation.

In view of VaR's deficiency, a coherent risk measure named Expected Shortfall (also called 'conditional VaR', 'means excess loss', 'beyond VaR' or 'tail VaR') is suggested by Artzner et al. [3] to complement VaR, which aims at measuring the risk of losses beyond VaR. It is defined as the conditional expectation of loss given that this loss is beyond the VaR.

Suppose X is a random variable denoting the loss of a given portfolio and $VaR_\alpha(X)$ is the VaR of the portfolio at the $100(1-\alpha)\%$ confidence level (i.e. the upper 100α percentile of the loss distribution), then

$$ES_\alpha(X) = E[-X| - X \geq VaR_\alpha(X)] \tag{6.1}$$

Here we assume that the loss distribution is continuous. If it is discrete, the definition of ES needs to be modified a little in order to make it a coherent risk measure, see Yamai and Yoshiba [11].

A number of comparative analyses on ES and VaR have been carried out by many researchers. See for example, Acerbi et al. [1], Acerbi and Tasche [2], Rockafeller and Uryasev [9], Tasche [10], Yamai and Yoshiba ([11], [12], [13], [14]). The advantages of ES over VaR include:

i) ES is sub-additive and therefore coherent.
ii) ES reflects the loss beyond VaR level and less likely to suggest perverse portfolio construction.
iii) ES reduces credit concentration.

iv) Adopting ES as a risk management tool is more conservative as more economic capital is required although the capital calculated is hard to interpret with respect to firms' default probability and does not necessarily correspond to the capital needed to maintain the firms' default probability below some specific level.

It is also known that under the assumption of normal return distribution with mean 0, expected shortfall provides equivalent risk management information to that of VaR since they are scalar multiple of each other.

Recently the finance community has shown great interests about credit risks. Jarrow et al. [6] propose the use of Markov chain model to incorporate the firms' credit rating in debt valuation. Based on this idea, Kijima and Komoribayashi [7] made some further studies while Arvanitis et al. [5] and Yang ([15], [16]) built credit spread models and ruin theory models, respectively.

In this chapter, we follow the idea in Yang [15] to present a model for measuring market and credit risks. In order to take into account the dependency of the credit risk, we propose to use a weak Markov chain rather than a Markov chain to model transition probabilities between credit states.

Let I_t be a time-homogeneous weak Markov chain of order r, $r \geq 1$, with finite state space $N = (1, 2, ..., k)$ representing k different credit states for $t \geq r - 1$. If $i_0, i_1, \ldots, i_{n+1} \in N$, then we have

$$\begin{aligned} &P\left[I_{n+1} = i_{n+1} | I_0 = i_0, I_1 = i_1, ..., I_{n-1} = i_{n-1}, I_n = i_n\right] \\ &= \left[I_{n+1} = i_{n+1} | I_{n-r+1} = i_{n-r+1}, ..., I_{n-1} = i_{n-1}, I_n = i_n\right] \end{aligned} \tag{6.2}$$

Expression (6.2) tells us that the probability of moving to state i_{n+1} given full histories of credit states is equivalent to the transition probability given only past r periods of histories.

Suppose the Markov chain is time-homogeneous. Then we can write

$$\begin{aligned} P\left[I_{n+1} = i_{n+1} | I_{n-r+1} = i_{n-r+1}, ..., I_{n-1} = i_{n-1}, I_n = i_n\right] \\ = q_{i_{n-r+1} i_{n-r+2} \cdots i_{n-1} i_n i_{n+1}} \end{aligned}$$

where $i_{n-r+1}, i_{n-r+2}, \ldots, i_{n+1} \in N$. We can construct the $k^r \times k$ transition matrix

$$\begin{bmatrix} q_{11...111} & q_{11...112} & \cdots & q_{11...11(k-1)} & q_{11...11k} \\ q_{11...121} & q_{11...122} & \cdots & q_{11...12(k-1)} & q_{11...12k} \\ \vdots & \vdots & & \vdots & \vdots \\ q_{11...1k1} & q_{11...1k2} & \cdots & q_{11...1k(k-1)} & q_{11...1kk} \\ q_{11...211} & q_{11...212} & \cdots & q_{11...21(k-1)} & q_{11...21k} \\ \vdots & \vdots & & \vdots & \vdots \\ q_{11...2k1} & q_{11...2k2} & \cdots & q_{11...2k(k-1)} & q_{11...2kk} \\ \vdots & \vdots & & \vdots & \vdots \\ q_{kk...kk1} & q_{kk...kk2} & \cdots & q_{kk...kk(k-1)} & q_{kk...kkk} \end{bmatrix}. \tag{6.3}$$

The objective of Yang [15] is to build a model that pools both market risks and credit risks together. Therefore, he first models credit risk ratings by a Markov chain, then constructs a surplus process as a function of portfolio returns, credit ratings and time. In addition, recursive equations for VaR calculations are obtained. For various portfolio returns assumption like normal distribution and shifted t-distribution, numerical illustrations are also provided.

In the next section, we will use the setup of Yang [15] and discuss the ES. We will present some numerical results to illustrate the ideas. In section 6.3, weak Markov chain will be used to model the credit ranking change. The final section summarizes the chapter.

6.2 Markov regime-switching model

In this section, we will apply the Markov chain model to represent the credit rating dynamics. We obtain the estimated credit rating transition probabilities based on historical data from available sources. In particular, the transition matrix is obtained by using the matrix given in JP Morgan [8] and conditional on the non-default states.

Let I_t be a time-homogeneous Markov chain with finite state space $N = (1, 2, ..., 7)$ representing 7 different credit states (non-default).

From (6.2) and (6.3), we have the transition probability for $r = 1$

$$\begin{aligned} &P\left[I_{n+1} = i_{n+1} | I_0 = i_0, I_1 = i_1, ..., I_{n-1} = i_{n-1}, I_n = i_n\right] \\ &= \left[I_{n+1} = i_{n+1} | I_n = i_n\right] = q_{i_n i_{n+1}} \end{aligned} \tag{6.4}$$

and the transition matrix of 7×7 is given by

$$Q = \begin{bmatrix} q_{11} & q_{12} & \cdots & q_{16} & q_{17} \\ q_{21} & q_{22} & \cdots & q_{26} & q_{27} \\ \vdots & \vdots & & \vdots & \vdots \\ q_{61} & q_{62} & \cdots & q_{66} & q_{67} \\ q_{71} & q_{72} & \cdots & q_{76} & q_{77} \end{bmatrix} = \begin{bmatrix} .9081 & .0833 & .0068 & .0006 & .0012 & .0000 & .0000 \\ .0070 & .9065 & .0779 & .0064 & .0006 & .0014 & .0002 \\ .0009 & .0227 & .9111 & .0552 & .0074 & .0026 & .0001 \\ .0002 & .0033 & .0596 & .8709 & .0531 & .0117 & .0012 \\ .0003 & .0014 & .0068 & .0781 & .8140 & .0893 & .0101 \\ .0000 & .0012 & .0025 & .0045 & .0684 & .8805 & .0429 \\ .0027 & .0000 & .0028 & .0162 & .0296 & .1401 & .8086 \end{bmatrix}, \tag{6.5}$$

where for $i = (1, 2, ..., 7)$, $\sum_{j=1}^{7} q_{ij} = 1$.

In this study, we adopt the notion from JP Morgan [8] in which various credit ratings were grouped into 7 categories. We regard state 1 as the highest credit class which correspond to Moody's Aaa or S&P's AAA while state 7 as the lowest, corresponding to Moody's Caa or S&P's CCC grade. With this setup, we can define the credit state dependent surplus process.

$$U_t = u + \sum_{m=1}^{t} X_m^{I_{m-1}} = u + \Delta Y_t, \tag{6.6}$$

where u is the initial surplus of the firm, X_m^i is the return in the mth time interval given that the firm's credit rating is of class i. δY_t refers to the aggregated return over entire time period. We assume that X_m^i, $i = 1, 2, ..., k$, $m = 1, 2, ...$ are independent random variables. We further assume that for any fixed $i = 1, 2..., k$, ΔX_m^i, $(m = 1, 2, ...,)$ are identically distributed and $\Delta X^i, ..., \Delta X^k$ are independent but not necessary follow the same distribution. Therefore, the portfolio return of a firm in each time interval depends only on the credit state at the start of each period but not on other random variables in the model.

For the surplus process defined in (6.6), let $T = inf\{n; U_n \leq 0\}$ be the default time of the firm. It is obvious that T is a stopping time. Then, we define the n-period (100-α)% VaR, if default does not occur before n, as

$$\begin{aligned} P\{\Delta Y_n \leq -y,\ T \geq n | I_0 = i_0, U_0 = u\} = P\{\Delta X_1^{i_0} + \cdots + \Delta X_n^{I_{n-1}} \leq -y, \\ \Delta X_1^{i_0} > -u, \ldots, \Delta X_1^{i_0} + \cdots + \Delta X_{n-1}^{I_{n-2}} > -u\} = \alpha\%. \end{aligned} \tag{6.7}$$

Denote the probability of equation (6.7) as $d_n^{i_0}(u, y)$, then we can calculate this in a recursive manner:

For $n \geq 2$,

$$\begin{aligned} d_n^{i_0}(u,y) &= P\{\Delta X_1^{i_0} + \cdots + \Delta X_n^{I_{n-1}} \leq -y, \\ &\quad \Delta X_1^{i_0} > -u, \ldots, \Delta X_1^{i_0} + \cdots + \Delta X_{n-1}^{I_{n-2}} > -u\} \\ &= \sum_{i=1}^{k} q_{i_0 i} \int_{-u}^{\infty} P\{\Delta X_2^i + \cdots + \Delta X_n^{I_{n-1}} \leq -(y+x), \\ &\quad \Delta X_2^i > -(u+x), \ldots, \Delta X_2^i + \cdots + \Delta X_{n-1}^{I_{n-2}} > -(u+x)\} f^{i_0}(x) dx \\ &= \sum_{i=1}^{k} q_{i_0 i} \int_{-u}^{\infty} d_{n-1}^i(u+x, y+x) f^{i_0}(x) dx, \end{aligned}$$

where $f^{i_0}(x)$ is the density function of $\Delta X_1^{i_0}$,

$$d_1^{i_0}(u,y) = P\{\Delta X_1^{i_0} \leq -y\}$$

and

$$d_2^{i_0}(u,y) = P\{\Delta X_1^{i_0} + \Delta X_2^{I_1} \leq -y, \Delta X_1^{i_0} > -u\}.$$

Note that it is possible that the VaR value, y, is larger than u. In this case, we say that the portfolio is very risky and some kind reconstruction of the firm's portfolio is required. In the case of default occurring before time n, practically it does not make much sense to calculate the VaR or ES for time period n.

For expected shortfall, let $F_{\Delta Y_n}(-y|I_0 = i_0)$ denote the distribution of ΔY_n given $T_0 = i_0$, we have

$$\begin{aligned} F_{\Delta Y_n}(-y|I_0 = i_0) &= P\{\Delta Y_n \leq -y \mid I_0 = i_0\} \\ &= P\{\Delta X_1^{i_0} + \cdots + \Delta X_n^{I_{n-1}} \leq -y\} \\ &= \sum_{i=1}^{k} q_{i_0 i} \int_{-\infty}^{\infty} P\{\Delta X_2^{i} + \cdots + \Delta X_n^{I_{n-1}} \leq -(y+x)\} f^{i_0}(x)dx \\ &= \sum_{i=1}^{k} q_{i_0 i} \int_{-\infty}^{\infty} F_{\Delta Y_{n-1}}(-(y+x) \mid I_0 = i_0) f^{i_0}(x)dx. \end{aligned}$$

Therefore, the expected shortfall is given by:

$$\begin{aligned} ES &= E\big[-\Delta Y_n \mid \Delta Y_n \leq -VaR\big] \\ &= -\frac{\int_{-\infty}^{-VaR} y \cdot f_{\Delta Y_n}(y|I_0 = i_0)dy}{\alpha}, \end{aligned} \tag{6.8}$$

where $f_{\Delta Y_n}(y|I_0 = i_0)$ is the density function of ΔY_n given $T_0 = i_0$.

An alternative way of calculating the VaR and the expected shortfall is to use Monte Carlo simulation. Monte Carlo method is a standard numerical tool in finance nowadays. In the following, we give an example to illustrate the idea. We use a shifted gamma distribution as the portfolio change distribution and conduct some numerical studies. We assume that the transition matrix for the credit risk rankings is given by (6.5) and the model parameters in the gamma distribution are set to reflect the credit risk and market risk. The density function of gamma distribution is

$$f(x) = \frac{(x/\theta)^{\beta} e^{-(x/\theta)}}{x\Gamma(\beta)}.$$

We assume that $X^{I_m = i} \sim (Ga(\alpha_i, \theta_i) - 4)$. Larger values of α are selected for the better credit states to reflect its relatively low tail risk, as well as a better return prospect resulted from a relative low credit costs. Smaller values of α are used for lower credit states to reflect the high credit risk. For simplicity, we fix $\theta = 1$. The values of parameter α are presented in Table 6.1 below.

Initial Credit State	1 2 3 4 5 6 7
α	7 6 5 4 3 2 1

Table 6.1. Parameters used for Gamma distribution return scenario

In Table 6.2 below, the initial credit state, estimated values, standard deviation and confidence intervals of VaR and ES are shown for $\alpha = 1\%$ and $n = 3$, and Table 6.3 shows the numerical results for $\alpha = 5\%$ and $n = 3$.

Initial Credit State	Risk Measure	Value	S.D.	95% CI
1	VaR	2.2269	0.1630	(1.9074, 2.5464)
	ES	4.0599	0.1918	(3.6839, 4.4358)
2	VaR	4.4014	0.1380	(4.1308, 4.6719)
	ES	6.1309	0.1531	(5.8308, 6.4311)
3	VaR	5.8705	0.1122	(5.6507, 6.0903)
	ES	7.0690	0.1294	(6.8154, 7.3226)
4	VaR	7.7040	0.0966	(7.5146, 7.8934)
	ES	8.6067	0.1032	(8.4044, 8.8091)
5	VaR	9.5340	0.0701	(9.3967, 9.6713)
	ES	10.1315	0.0678	(9.9986, 10.2645)
6	VaR	10.5606	0.0461	(10.4702, 10.6509)
	ES	10.9233	0.0460	(10.8332, 11.0134)
7	VaR	11.4469	0.0204	(11.4070, 11.4869)
	ES	11.5950	0.0189	(11.5580, 11.6320)

Table 6.2. Simulated VaR and ES with $\alpha = 1\%$ and $n = 3$

Initial Credit State	Risk Measure	Value	S.D.	95% CI
1	VaR	-0.8778	0.0777	(-1.0302, -0.7254)
	ES	1.0371	0.0896	(0.8615, 1.2127)
2	VaR	1.5097	0.0711	(1.3704, 1.6491)
	ES	3.2981	0.0738	(3.1535, 3.4428)
3	VaR	3.5607	0.0616	(3.4400, 3.6814)
	ES	4.9771	0.0687	(4.8424, 5.1118)
4	VaR	5.7474	0.0557	(5.6383, 5.8565)
	ES	6.9356	0.0.0598	(6.8184, 7.0528)
5	VaR	8.0399	0.0451	(7.9515, 8.1284)
	ES	8.9426	0.0448	(8.8547, 9.0304)
6	VaR	9.5806	0.0322	(9.5175, 9.6437)
	ES	10.1723	0.0321	(10.1094, 10.2352)
7	VaR	10.9434	0.0185	(10.9072, 10.9796)
	ES	11.2482	0.0156	(11.2175, 11.2788)

Table 6.3. Simulated VaR and ES with $\alpha = 5\%$ and $n = 3$

The numerical results in Tables 6.2 and 6.3 are consistent with our intuition and the definitions of VaR and expected shortfall. The numerical results show that ES is always larger than the corresponding VaR. The values of VaR and ES for $\alpha = 1\%$ are larger than those corresponding VaR and ES for $\alpha = 5\%$. In case the ES is larger than u, that means the firm has to do something with its portfolio.

6.3 Weak Markov-regime switching model

In this section, we construct a model of market and credit risks with credit transition being described by a weak Markov chain. As we know dependent structure is a common phenomenon in finance and is difficult to deal with. To demonstrate the idea, we only use a second order weak Markov chain here. That is, we assume $r = 2$ in (6.3).

As in section 6.2, we would attempt to formulate the surplus process with second-order Markov regime-switching instead of the first order. By referring to Yang's (2000, 2003) model, we can define an analogous surplus process as

$$U_t = u + \sum_{m=1}^{t} \Delta X_m^{I_{m-2} I_{m-1}} = u + \Delta Y_t$$

Again, let T be the default time (i.e. $T = inf\{t; U_t \leq 0\}$), $U_0 = u$ be the initial surplus and further assume that $I_0 = i_0$ and $I_{-1} = i_{-1}$ are known credit ratings in current and last periods respectively, then we can obtain an iterative formula in computing n-period VaR by considering

$$P\{\Delta Y_n \leq -y, T \geq n \,|\, I_0 = i_0, I_{-1} = i_{-1}, U_0 = u\} = \alpha. \tag{6.9}$$

The derivation of a recursive formula in the calculation of VaR is similar to that in section 6.2. Furthermore, if we treat

$$\{(1,2),(1,3),\cdots,(1,k),(2,1),\cdots,(2,k),\cdots,(k,1),\cdots(k,k)\}$$

as the state space, then we can construct a new Markov chain from the weak Markov chain. After constructing the new Markov chain, mathematically, the problem reduces to that in section 6.2; the only difference is that the state space of the new Markov chain has $k \times k$ states. Therefore we can also calculate the VaR by using the formulas in section 6.2.

Similar to equation (6.8), we have the following expression for the expected shortfall.

$$\begin{aligned} ES &= E\big[-\Delta Y_n \,\big|\, \Delta Y_n \leq -VaR\big] \\ &= -\frac{\int_{-\infty}^{-VaR} y \cdot f_{\Delta Y_n}(y \,|\, I_{-1} = i_{-1}, I_0 = i_0) dy}{\alpha}. \end{aligned}$$

Again, $f_{\Delta Y_n}(y \mid I_{-1} = i_{-1}, I_0 = i_0)$ is the density function of ΔY_n given $I_{-1} = i_{-1}$ and $I_0 = i_0$, and can be obtained either by a similar recursive formula to that in section 6.2 or, by restructuring the larger state space and reducing the second-order Markov chain to a first-order one, then use the formula in section 6.2.

6.4 Concluding remarks

In this chapter, we have presented a model which can incorporate both market and credit risks. Our model can also deal with some kind of dependency structure of the credit risk. We provide two ways of calculating the VaR and ES; one is by using the recursive equations developed in this chapter and the other is through the use of standard Monte Carlo method. Both methods are computationally intensive However, with today's powerful computer, hopefully this will not be a big hurdle.

The risk measures for portfolio with both market and credit risks are practically important and theoretically interesting. This chapter provides simple models that both handle these risks.

References

1. Acerbi, C., Nordio, C. and C. Sirtori (2001). "Expected shortfall as a tool for financial risk management," Working paper, Italian Association for Financial Risk Management.
2. Acerbi, C. and D. Tasche (2002). "On the coherence of expected shortfall," *Journal of Banking and Finance*, 26(7): 1487–1503.
3. Artzner, P., Delbaen, F., Eber, J. M. and D. Heath (1997). "Thinking coherently," *Risk*, 10(11): 68–71.
4. Artzner, P., Delbaen, F., Eber, J. M. and D. Heath (1999). "Coherent measures of risk," *Mathematical Finance*, 9(3): 203–228.
5. Arvanitis, A., Gregory, J. and J.P. Laurent (1999). "Building models for credit spreads," *Journal of Derivatives*, 6: 27–43.
6. Jarrow, R. A., Lando, D. and S.M. Turnbull (1997). "A Markov model for the term structure of credit risk spread," *Review of Financial Studies*, 10: 481–523.
7. Kijima, M. and K. Komoribayashi (1998). "A Markov chain model for valuing credit risk derivatives," *Journal of Derivatives*, 5: 97–108.
8. Morgan, J. P. (1997). *Introduction to CreditMetrics*, J. P. Morgan and Company, New York.
9. Rockafellar, R. T. and S. Uryasev (2001). "Conditional value-at-risk for general loss distributions," *Journal of Banking and Finance*, 26(7): 1443–1471.
10. Tasche, D. (2002). "Expected shortfall and beyond," *Journal of Banking and Finance*, 26: 1519–1533.

11. Yamai, Y. and T. Yoshiba (2002a), "On the validity of value-at-risk: comparative analyses with expected shortfall," *Institute of Monetary and Economic Studies*, 20(1), Bank of Japan, 57–86.
12. Yamai, Y. and T. Yoshiba (2002b). "Comparative analyses of expected shortfall and value-at-risk: their estimation error, decomposition and optimization," *Institute of Monetary and Economic Studies*, 20(1), Bank of Japan, 87–122.
13. Yamai, Y. and T. Yoshiba (2002c). "Comparative analyses of expected shortfall and value-at-risk (2): expected utility maximization and tail risk," *Institute of Monetary and Economic Studies*, 20(2), Bank of Japan, 95–115.
14. Yamai, Y. and T. Yoshiba (2002d). "Comparative analyses of expected shortfall and value at risk (3): their validity under market stress," *Institute of Monetary and Economic Studies*, 20(3), Bank of Japan, 181–238.
15. Yang, H. (2000). "An integrated risk management method: VaR approach," *Multinational Finance Journal*, 4(3,4): 201–219.
16. Yang, H. (2003). "Ruin theory in a financial corporation model with credit risk," *Insurance: Mathematics and Economics*. 33: 135–145.

7

Filtering of Hidden Weak Markov Chain -Discrete Range Observations

Shangzhen Luo[1] and Allanus H. Tsoi[2]

[1] Department of Mathematics
University of Northern Iowa
Cedar Falls, IA 50614-0506
USA
luos@uni.edu

[2] Department of Mathematics
University of Missouri
Columbia, Missouri 65211
USA
tsoi@math.missouri.edu

Summary. In this paper we consider a hidden discrete time finite state process X whose behavior at the present time t depends on its behavior at the previous k time steps, which is a generalization of the usual hidden finite state Markov chain, in which k equals to one. We consider the case when the range space of our observations is finite. We present filtering equations for certain functionals of the chain and perform related error analysis.

Key words: Hidden weak Markov chain, filtering, smoothing, EM algorithm, parameter reestimation.

7.1 Introduction

Stochastic filtering of a hidden Markov chain constitutes a large volume of literature. See for example, the survey paper [11] by Ephraim and Merhav, which discusses both the theoretical and applied aspect of this topic. If we associate the state space of the Markov chain, which represents the signal, as well as that of the observations, with the Euclidean standard unit vectors in some finite dimensional Euclidean spaces $\mathbb{R}^m$ and $\mathbb{R}^n$, then it turns out that the calculations involved are quite simplified. This has been the approach by Elliott et. al. [5]. In the past few years some applications of hidden Markov modeling in financial problems have appeared (e.g.,[6],[7] by Elliott et. al.).

However, there have been conclusions that a quite significant number of financial entities, which, when being modeled by stochastic evolution equations, possess memories. This phenomenon is also called the Joseph effect (see [2] by Cutland, Kopp and Willinger). Therefore, if we want to estimate some functionals of those financial entities which cannot be observed directly, through stochastic filtering techniques, then we have to develop some adequate filtering theory for the case when the signals possess memories. This is the main motivation of writing this paper. Here we consider a discrete time finite state chain X, whose behavior at the present time t depends on its behavior in the most recent k time steps. See the definition in Section 2. We call such a process a weak Markov chain of (memory)order k. This should not be confused with the usual definition of the order of a Markov chain, which corresponds to the dimension of the state space. This kind of weak Markov chain has been dealt with by Wang [22], but it has not been that popular.

One possible way to construct filtering equations for the weak Markov chain which represents the signal is to Markovianize the weak chain to turn it to an ordinary Markov chain, and then apply the classical methods to the Markovianized chain. We are assuming that the transition probabilities of our chain to be independent of time. That is, the chain is homogeneous with respect to the time parameter. If the dimension of the state space of our original weak chain (X_n) is N, and if we denote our Markovianized chain by (Y_n), then we have

$$Y_n = (X_{n-d+1}, \dots, X_n),$$

where $d > 1$ is the memory order of (X_n). Consequently, the dimension of the transition matrix of (Y_n) would become $N^d \times N^d$. On the other hand, if we employ our proposed method in this paper with the help of our function α given in Proposition 2.1, part (iii) in Section 4, we only need to consider a transition matrix with dimension $N \times N^d$. Therefore, from the numerical computational point of view, our approach would be more efficient than the method of Markovianizing the original chain (X_n), especially when the memory order d and the dimension of the state space N are very large.

The traditional methodology for the dynamical system approach of deriving filtering, smoothing and prediction recursive equations for functionals of a hidden Markov chain with finite state space and of finite dimension N is to assume the state space of the chain to consist of the N canonical unit vectors $e_1, e_2, \dots, e_N$, where e_i is the unit vector with the value 1 in the ith co-ordinate and 0 elsewhere. The procedures and calculations involved rely heavily on this assumption. (cf. Elliott et al. [5]) In the case of our weak Markov chain with memory order $d > 1$, if we simply perform the Markovianization, that is, if we simply consider the new Markov chain (Y_n) given by

$$Y_n = (X_{n-d+1}, \dots, X_n),$$

then the set of possible values which Y_n assume would no longer have the form of the canonical unit vectors. As a result, we will still need to construct a mapping like α given on page 3 to turn the Y_n's into canonical unit vectors before we can perform calculations in the traditional way. Hence the methodology we present in this paper possesses more advantage than the method of simply Markovianizing the weak Markov chain.

The paper is organized as follows Section 2 presents with the basic definitions, settings, and preliminary results of our filtering problem. In Section 3 we consider the issue on measure change. We give a general un normalized recursive filter in Section 4. The proofs of most of the theorems, propositions and lemmas stated in Section 2, 3 and 4 are direct generalizations of those described in references [5], [9] and [10]. Therefore we just state these theorems, propositions and lemmas and refer the readers to these references. In Section 5 we give filter estimates for the states of the process, the number of jumps from one particular state to another specified state in a certain fixed time interval, and occupation times. Finally we consider parameter re-estimation in Section 6 and error analysis for our filters in Section 7. The paper ends with a Conclusion Section.

7.2 Basic Settings

We suppose all random variables are defined on a probability space $(\Omega, \mathcal{F}, P)$. We say that $X = \{X_n\}_{n\geq 0}$ is a weak Markov chain of (memory) order k, $k \geq 1$, with finite state space S_X if for $n \geq k-1$, $x_0, ..., x_n, x_{n+1} \in S_X$, we have

$$\begin{aligned} &P(X_{n+1} = x_{n+1}|X_0 = x_0, X_1 = x_1, ..., X_{n-1} = x_{n-1}, X_n = x_n) \\ &= P(X_{n+1} = x_{n+1}|X_{n-k+1} = x_{n-k+1}, ..., X_{n-1} = x_{n-1}, X_n = x_n). \end{aligned} \tag{7.1}$$

Throughout this paper, in order to avoid unnecessary complicated notation, we simply consider a weak Markov chain of order 2. The results in this paper can readily be extended to weak Markov chains of order k, for $k \geq 1$. See also the concluding remark in the last section. From now on $\{X_k\}_{k\geq 0}$ is a weak Markov chain of order 2 and with state space $S_X = \{e_1, e_2, ..., e_N\} \subset \mathbb{R}^N$ which is the collection of N-dimension standard unit vectors (see [5]).

Thus we have, for $k \geq 1$,

$$\begin{aligned} &P(X_{k+1} = x_{k+1} \mid X_0 = x_0, X_1 = x_1, ..., X_{k-1} = x_{k-1}, X_k = x_k) \\ &= P(X_{k+1} = x_{k+1} \mid X_{k-1} = x_{k-1}, X_k = x_k). \end{aligned} \tag{7.2}$$

Assume that the chain is homogeneous, so that we can write

$$P(X_{k+1} = e_i \mid X_k = e_j, X_{k-1} = e_l) = a_{ijl},$$

where $i, j, l \in \{1, 2, ..., N\}$. Then we have the $N \times N^2$ transition matrix

$$A = \begin{pmatrix} a_{111} & a_{112} & \dots & a_{11N} & \dots & a_{1N1} & a_{1N2} & \dots & a_{1NN} \\ a_{211} & a_{212} & \dots & a_{21N} & \dots & a_{2N1} & a_{2N2} & \dots & a_{2NN} \\ & & \dots & & & & \dots & & \\ a_{N11} & a_{N12} & \dots & a_{N1N} & \dots & a_{NN1} & a_{NN2} & \dots & a_{NNN} \end{pmatrix}.$$

Suppose $\{T_k\}_{k\geq 1}$ is a sequence of observations of the form

$$T_{k+1} = c(X_k, \omega_{k+1}),$$

where $\{\omega_k\}_{k\geq 1}$ is an independent identically distributed (IID) sequence of noise which is independent of $\{X_k\}_{k\geq 0}$, and $c(\cdot,\cdot)$ is a deterministic Borel measurable function. Assume that the observations take values in the set $S_T = \{f_1, f_2, ..., f_M\} \subset \mathbb{R}^M$ which is the collection of M-dimensional standard unit vectors. Write $\mathcal{F}_k := \sigma(X_0, X_1, ..., X_k)$, $\mathcal{G}_k := \sigma(X_0, X_1, ..., X_k, T_1, ..., T_k)$, and $\mathcal{Y}_k := \sigma(T_0, T_1, ..., T_k)$.

Proposition 1.

(*i*) $P(X_{k+1} = x_{k+1} \mid X_0 = x_0, ..., X_k = x_k, T_1 = t_1, ..., T_k = t_k)$
$= P(X_{k+1} = x_{k+1} \mid X_{k-1} = x_{k-1}, X_k = x_k) \quad for \quad k \geq 1;$

(*ii*) $P(T_{k+1} = t_{k+1} \mid X_0 = x_0, ..., X_k = x_k, T_1 = t_1, ..., T_k = t_k)$
$= P(T_{k+1} = t_{k+1} \mid X_k = x_k) \quad for \quad k \geq 0;$

(*iii*) $E(X_{k+1} \mid \mathcal{G}_k) = A\alpha \begin{pmatrix} X_k \\ X_{k-1} \end{pmatrix} = E(X_{k+1} \mid X_k, X_{k-1}) \quad for \quad k \geq 1;$

(*iv*) $E(T_{k+1} \mid \mathcal{G}_k) = CX_k = E(T_{k+1} \mid X_k) \quad for \quad k \geq 0.$

where α is a map defined by: $\alpha \begin{pmatrix} e_r \\ e_s \end{pmatrix} = e_{rs}$ *with* $e_{rs} := (0, \dots, 0, 1, 0, \dots, 0)' \in \mathbb{R}^{N^2}$, *for* $1 \leq r, s \leq N$, *where* 1 *is at the* $((r-1)N + s)^{th}$ *position, and the prime* $'$ *denotes the transpose. The matrix* C *above is*

$$C = \begin{pmatrix} c_{11} & c_{12} & \dots & c_{1N} \\ c_{21} & c_{22} & \dots & c_{2N} \\ & & \dots & \\ c_{M1} & c_{M2} & \dots & c_{MN} \end{pmatrix},$$

where $c_{ij} = P(T_{k+1} = f_i | X_k = e_j), \quad 1 \leq j \leq N, \quad 1 \leq i \leq M$. *The matrix* C *is called the state to observation transition matrix.*

Proof. By the weak Markov property and because of the independence of $\{\omega_k\}_{k=1}^{\infty}$ and X, we obtain (i). By independence we also obtain (ii). Part (iii) follows from part (i) and from the property of the conditional expectation operator. Part (iv) follows from part (ii).

Next define

$$V_{k+1} := X_{k+1} - A\alpha\begin{pmatrix} X_k \\ X_{k-1} \end{pmatrix}, \quad k \geq 1;$$
$$W_{k+1} := T_{k+1} - CX_k, \quad k \geq 0.$$

We note from Proposition 1 part *(iii)* that $\{V_{k+1}\}$ is a sequence of martingale increments with respect to the filtration $\{\mathcal{G}_k\}$, so that is the dynamic representation for the weak Markov Chain X

$$X_{k+1} = A\alpha\begin{pmatrix} X_k \\ X_{k-1} \end{pmatrix} + V_{k+1}, \quad k \geq 1. \tag{7.3}$$

Next we state the following lemma:

Lemma 1. *For* $k \geq 1$,

$$\begin{aligned} (i) \quad V_{k+1}V'_{k+1} &= \text{diag}(A\alpha\begin{pmatrix} X_k \\ X_{k-1} \end{pmatrix}) + \text{diag}(V_{k+1}) - A\text{diag}(\alpha\begin{pmatrix} X_k \\ X_{k-1} \end{pmatrix})A' \\ &\quad - A\alpha\begin{pmatrix} X_k \\ X_{k-1} \end{pmatrix}V'_{k+1} - V_{k+1}(A\alpha\begin{pmatrix} X_k \\ X_{k-1} \end{pmatrix})'; \\ (ii) \quad < V_{k+1} >: &= E(V_{k+1}V'_{k+1} \mid \mathcal{F}_k); \\ &= E(V_{k+1}V'_{k+1} \mid X_k, X_{k-1}) \\ &= \text{diag}(A\alpha\begin{pmatrix} X_k \\ X_{k-1} \end{pmatrix}) - A\text{diag}(\alpha\begin{pmatrix} X_k \\ X_{k-1} \end{pmatrix})A'. \end{aligned}$$

Proof. See [5].

7.3 Change of Measure

Write $T_k^i := \langle T_k, f_i \rangle$, $c_{k+1} = E(T_{k+1} \mid \mathcal{G}_k) = CX_k$. Hence $c_{k+1}^i := \langle c_{k+1}, f_i \rangle (= \langle CX_k, f_i \rangle = \sum_j c_{ij}\langle X_k, e_j \rangle = \Pi_j c_{ij}^{\langle X_k, e_j \rangle})$, for $1 \leq i \leq M$ and $1 \leq j \leq N$. Define

$$\lambda_l = \prod_{i=1}^{M}\left(\frac{1}{Mc_l^i}\right)^{T_l^i}, \quad \Lambda_k = \prod_{l=1}^{k}\lambda_l,$$

where $l, k \in \mathbb{N}$. Note that $\lambda_l = \sum_{i=1}^{M}\frac{T_l^i}{Mc_l^i}$. Consequently, we have

Lemma 2.

$$E(\lambda_{k+1} \mid \mathcal{G}_k) = 1. \tag{7.4}$$

Now define a new probability measure $\bar{P}$ on $\left(\Omega, \bigvee_{l=1}^{\infty} \mathcal{G}_l\right)$ by:

$$\left.\frac{d\bar{P}}{dP}\right|_{\mathcal{G}_k} = \Lambda_k.$$

The existence of $\bar{P}$ follows from the Kolmogorov's extension theorem.

Theorem 1. (Conditional Bayes' Theorem) *Given a probability space $(\Omega, \mathcal{F}, P)$, and $\mathcal{G} \subseteq \mathcal{F}$ is a sub-σ-field. Suppose that $\bar{P}$ is another probability measure which is abslolutely continuous with respect to P and with Radon-Nikodym derivative $\frac{d\bar{P}}{dP} = \Lambda$. Then for any integrable $\mathcal{F}$-measurable r.v. ϕ,*

$$\bar{E}(\phi \mid \mathcal{G}) = \frac{E(\Lambda\phi \mid \mathcal{G})}{E(\Lambda \mid \mathcal{G})}.$$

Consequently we have

Lemma 3. *If $\{\phi_k\}$ is a $\mathcal{G} = \{\mathcal{G}_k\}$ adapted integrable sequence of r.v.'s, then*

$$\bar{E}(\phi_k \mid \mathcal{Y}_k) = \frac{E(\Lambda_k \phi_k \mid \mathcal{Y}_k)}{E(\Lambda_k \mid \mathcal{Y}_k)}.$$

In addition, we have the following (see [5])

Lemma 4. *$\bar{P}(T_{k+1}^j = 1 \mid \mathcal{G}_k) = \frac{1}{M}$, so that under $\bar{P}$, T_{k+1} is independent of $\mathcal{G}_k$ and $\{T_k\}$, and is an IID. sequence of uniformly distributed random variables, with $\bar{P}(T_k = f_j) = \frac{1}{M}$, for $1 \leq j \leq M$ and $\forall k \in \mathbb{N}$.*

More generally, we have $E(T_{k+1}^j \mid \mathcal{G}_k, X_{k+1}) = c_{k+1}^j$. Also we have $\bar{P}(T_{k+1}^j = 1 | \mathcal{G}_k, X_{k+1}) = \frac{1}{M}$ which implies that T_{k+1} is independent of $\mathcal{G}_k$ and X_{k+1}. Moreover, we have the following theorem:

Theorem 2. $\bar{E}(X_{k+1} \mid \mathcal{G}_k) = E(X_{k+1} \mid \mathcal{G}_k) = A\alpha \begin{pmatrix} X_k \\ X_{k-1} \end{pmatrix}.$

For the proof see reference [5].

Next we consider a reverse measure change, starting with a probability measure $\bar{P}$ on $(\Omega, \bigvee_{l=1}^{\infty} \mathcal{G}_l)$ equipped with the following:
1. The process X is a finite-state weak Markov chain with transition matrix A and has dynamics given by (7.3);
2. $\{T_k\}_{k=1}^{\infty}$ is a sequence of IID. random variables and $\bar{P}(T_{k+1}^i = 1 | \mathcal{G}_k, X_{k+1}) = \frac{1}{M}$.

Suppose $C = (c_{ji})$,$1 \leq j \leq M$, $1 \leq i \leq N$; $\sum_{j=1}^{M} c_{ji} = 1$, $c_{ji} \geq 0$. Construct P from $\bar{P}$ by setting

$$\bar{\lambda}_l = \prod_{i=1}^{M} (Mc_l^i)^{T_l^i}, \quad \bar{\Lambda}_k = \prod_{l=1}^{k} \bar{\lambda}_l,$$

where $l, k \in \mathbb{N}$ and $c_l^i = \langle CX_{l-1}, f_i \rangle$.
As in the previous case, we have $\bar{E}(\bar{\lambda}_{k+1} \mid \mathcal{G}_k) = 1$.
In addition, we can show that $\bar{E}(\bar{\lambda}_{k+1} \mid \mathcal{G}_k, X_{k+1}) = 1$. The Radon-Nikodym derivative is defined by

$$\left.\frac{dP}{d\bar{P}}\right|_{\mathcal{G}_k} = \bar{\Lambda}_k.$$

Consequently, that we have the following two lemmas:

Lemma 5. $E(X_{k+1} \mid \mathcal{G}_k) = \bar{E}(X_{k+1} \mid \mathcal{G}_k) = A\alpha \begin{pmatrix} X_k \\ X_{k-1} \end{pmatrix}$.

Lemma 6. $E(T_{k+1} | \mathcal{G}_k) = CX_k$.

7.4 A general unnormalized recursive filter

Notation 7.4.1 *For any process $\{H_k\}$, $k \in \mathbb{N}$, we write*

$$\gamma_k(H_k) := \bar{E}(\bar{\Lambda}_k H_k \mid \mathcal{Y}_k), \quad \hat{H}_k = E(H_k \mid \mathcal{Y}_k).$$

By Bayes' theorem, we have

$$\hat{H}_k = E(H_k \mid \mathcal{Y}_k) = \frac{\bar{E}(\bar{\Lambda}_k H_k \mid \mathcal{Y}_k)}{\bar{E}(\bar{\Lambda}_k \mid \mathcal{Y}_k)} = \frac{\gamma_k(H_k)}{\gamma_k(1)}.$$

For $1 \leq l \leq N$, write

$$c_l(T_{k+1}) = M \prod_{i=1}^{M} c_{il}^{T_{k+1}^i}.$$

Proposition 2. *For any random variable H, we have*

$$\bar{E}(\bar{\Lambda}_{k+1}H \mid \mathcal{Y}_{k+1}) = \sum_l c_l(T_{k+1})\bar{E}(\bar{\Lambda}_k H\langle X_k, e_l\rangle \mid \mathcal{Y}_{k+1}).$$

The proof is similar to the one given in reference [5].

Proposition 3. *For every $L_k \in \sigma(\mathcal{G}_k, \mathcal{Y}_{k+1})$, $\bar{E}(L_k V_{k+1}|\mathcal{G}_k, \mathcal{Y}_{k+1}) = 0$, so that*

$$\bar{E}(L_k V_{k+1}|\mathcal{Y}_{k+1}) = \bar{E}(\bar{E}(L_k V_{k+1}|\mathcal{G}_k, \mathcal{Y}_{k+1})|\mathcal{Y}_{k+1}) = 0\ .$$

Next we note that $\langle \alpha \begin{pmatrix} X_k \\ X_{k-1} \end{pmatrix}, e_{ls}\rangle = \langle X_k, e_r\rangle\langle X_{k-1}, e_s\rangle$.

Write

$$\begin{aligned} q_k(e_{ls}) &= \bar{E}\Big(\bar{\Lambda}_k \langle \alpha \begin{pmatrix} X_k \\ X_{k-1} \end{pmatrix}, e_{ls}\rangle \;\Big|\; \mathcal{Y}_k\Big) \\ &= \bar{E}(\bar{\Lambda}_k \langle X_k, e_l\rangle\langle X_{k-1}, e_s\rangle \mid \mathcal{Y}_k) \end{aligned}$$

and

$$q'_k(e_r) = \bar{E}(\bar{\Lambda}_k\langle X_k, e_r\rangle \mid \mathcal{Y}_k).$$

where $1 \le l, s, r, \le N$, $k \in \mathbb{N}$, $e_{ls} = \alpha \begin{pmatrix} e_l \\ e_s \end{pmatrix}$. Denote

$$p_k(e_r) = E(\langle X_k, e_r\rangle \mid \mathcal{Y}_k).$$

Then the following result is immediate.

Theorem 3. $\sum_s q_k(e_{ls}) = q'_k(e_l)$; $p_k(e_r) = \dfrac{\sum_s q_k(e_{rs})}{\sum_{r,s} q_k(e_{rs})}$.

Definition 1. *From now on, we assume*

$$H_{k+1} := \sum_{l=1}^{k+1}(\alpha_l + \langle \beta_l, V_l\rangle + \langle \delta_l, T_l\rangle),$$

Then we have the following recursive form of $\{H_k\}$:

$$H_{k+1} = H_k + \alpha_{k+1} + \langle \beta_{k+1}, V_{k+1}\rangle + \langle \delta_{k+1}, T_{k+1}\rangle,$$

where α_l is one dimension, β_l is N-dimension, δ_l is M-dimension and they are $\mathcal{G}_{l-1}$-measurable.

Notation 7.4.2 *For any $\mathcal{G}$-adapted process $\phi_k, k \in \mathbb{N}$, write*

$$\gamma_{m,k}(\phi_m) = \bar{E}\Big(\bar{\Lambda}_k \phi_m \alpha \begin{pmatrix} X_k \\ X_{k-1} \end{pmatrix} \Big| \mathcal{Y}_k\Big) \ ,$$
$$\gamma_{m,k}^{l,s}(\phi_m) = \langle \gamma_{m,k}(\phi_m), e_{ls} \rangle = \bar{E}\Big(\bar{\Lambda}_k \phi_m \Big\langle \alpha \begin{pmatrix} X_k \\ X_{k-1} \end{pmatrix}, e_{ls} \Big\rangle \Big| \mathcal{Y}_k\Big) \ .$$

As in [5], we have:

Theorem 4.

$$\gamma_{k+1,k+1}^{ls}(H_{k+1}) = c_s(T_{k+1}) \sum_v a_{lsv} \big[\gamma_{k,k}^{sv}(H_k + \alpha_{k+1} + \beta_{k+1}^l - \langle \beta_{k+1}, a_{.sv} \rangle) + \langle \gamma_{k,k}^{sv}(\delta_{k+1}), T_{k+1} \rangle\big],$$

where $a_{.sv} = Ae_{sv}$.

7.5 Estimation of states, transitions and occupation times

7.5.1 State estimation

Take $H_{k+1} = H_k = ... = H_0 = 1$, $\alpha_l = 0$, $\beta_l = 0$, $\delta_l = 0$, and by Theorem 4, we have the following recursive equation for $\{q_k\}$,

$$\begin{aligned} \gamma_{k+1,k+1}^{ls}(1) &= q_{k+1}(e_{ls}) \\ &= c_s(T_{k+1}) \sum_v a_{lsv} q_k(e_{sv}). \end{aligned} \tag{7.5}$$

Take $H_{k+1} = H_k = ... = H_n =< X_n, e_p >$ for $k+1 > n$. Then $\alpha_l = 0$, $\beta_l = 0$, $\delta_l = 0$ for $k+1 \geq l > n$. By applying Theorem 4 we obtain the following smoother.

$$\gamma_{k+1,k+1}^{ls}(\langle X_n, e_p \rangle) = c_s(T_{k+1}) \sum_v a_{lsv} \gamma_{n,k}^{sv}(\langle X_n, e_p \rangle).$$

7.5.2 Estimators for the number of jumps

The number of jumps from state (e_r, e_s) to e_t up to time k is given by

$$
\begin{aligned}
\mathcal{J}_k^{trs} &= \sum_{l=2}^{k} \langle X_l, e_t\rangle\langle X_{l-1}, e_r\rangle\langle X_{l-2}, e_s\rangle \\
&= \mathcal{J}_{k-1}^{trs} + \langle X_k, e_t\rangle\langle X_{k-1}, e_r\rangle\langle X_{k-2}, e_s\rangle \\
&= \mathcal{J}_{k-1}^{trs} + \left\langle A\alpha \begin{pmatrix} X_{k-1} \\ X_{k-2} \end{pmatrix}, e_t \right\rangle \langle X_{k-1}, e_r\rangle\langle X_{k-2}, e_s\rangle \\
&\quad + \langle V_k, e_t\rangle\langle X_{k-1}, e_r\rangle\langle X_{k-2}, e_s\rangle \\
&= \mathcal{J}_{k-1}^{trs} + a_{trs}\langle X_{k-1}, e_r\rangle\langle X_{k-2}, e_s\rangle \\
&\quad + \langle V_k, e_t\rangle\langle X_{k-1}, e_r\rangle\langle X_{k-2}, e_s\rangle.
\end{aligned}
$$

Take $H_{k+1} = \mathcal{J}_{k+1}^{trs}$, so that $\alpha_l = a_{trs}\langle X_{l-1}, e_r\rangle\langle X_{l-2}, e_s\rangle$, $\beta_l = e_t\langle X_{l-1}, e_r\rangle \times \langle X_{l-2}, e_s\rangle$, and $\delta_l = 0$. Thus

$$
\begin{aligned}
&\gamma_{k+1,k+1}^{lm}(\mathcal{J}_{k+1}^{trs}) \\
&= c_m(T_{k+1}) \sum_v a_{lmv} \gamma_{k,k}^{mv} \left[\mathcal{J}_k^{trs} + \langle X_k, e_r\rangle\langle X_{k-1}, e_s\rangle \right. \\
&\quad \left. \times (a_{trs} + \sigma_{tl} - a_{tmv}) \right] \\
&= c_m(T_{k+1}) \sum_v a_{lmv} \gamma_{k,k}^{mv}(\mathcal{J}_k^{trs}) + c_m(T_{k+1}) a_{lms} \sigma_{mr} \sigma_{lt} q_k(e_{ms}),
\end{aligned}
$$

where $\sigma_{ij} = \begin{cases} 1, & if\ i = j; \\ 0, & if\ i \neq j. \end{cases}$

Now take $H_{k+1} = H_k = ... = H_n = \mathcal{J}_n^{trs}$, where $k+1 > n$, so that $\alpha_l = 0$, $\beta_l = 0$, and $\delta_l = 0$ for $k+1 \geq l > n$.
Then we get

$$
\gamma_{n,k+1}^{lm}(\mathcal{J}_n^{trs}) = c_m(T_{k+1}) \sum_v a_{lmv} \gamma_{n,k}^{sv}(\mathcal{J}_n^{trs}).
$$

7.5.3 Estimators for 1-state occupation times

The number of occupations up to time k where the chain X was in state e_r is given by

$$
\mathcal{O}_k^r = \sum_{l=1}^{k} \langle X_{l-1}, e_r\rangle.
$$

Take $H_{k+1} = \mathcal{O}_{k+1}^r$, $\alpha_l = \langle X_{l-1}, e_r\rangle$, $\beta_l = 0$, $\delta_l = 0$. Then the filter equation is given by

$$
\begin{aligned}
\gamma_{k+1,k+1}^{lm}(\mathcal{O}_{k+1}^r) &= c_m(T_{k+1}) \sum_v a_{lmv} \left(\gamma_{k,k}^{mv}(\mathcal{O}_k^r) + \gamma_{k,k}^{mv}(\langle X_k, e_r\rangle) \right) \\
&= c_m(T_{k+1}) \sum_v a_{lmv} \left(\gamma_{k,k}^{mv}(\mathcal{O}_k^r) + \sigma_{mr} q_k(e_{mv}) \right).
\end{aligned}
$$

Let $H_{k+1} = H_k = ... = H_n = \mathcal{O}_n^r$ for $k+1 > n$, so that $\alpha_l = 0$, $\beta_l = 0$, $\delta_l = 0$ for $k+1 \geq l > n$. Then we obtain the smoother

$$\gamma_{n,k+1}^{lm}(\mathcal{O}_n^r) = c_m(T_{k+1}) \sum_v a_{lmv} \gamma_{n,k}^{lm}(\mathcal{O}_n^r).$$

7.5.4 Estimators for 2-state occupation times

The number of occupations up to time k where the weak Markov chain X was in state (e_r, e_s) is given by

$$\mathcal{O}_k^{rs} = \sum_{l=2}^{k} \langle X_{l-1}, e_r \rangle \langle X_{l-2}, e_s \rangle.$$

Take $H_k = \mathcal{O}_k^{rs}$, $\alpha_l = \langle X_{l-1}, e_r \rangle \langle X_{l-2}, e_s \rangle$, $\beta_l = 0$, $\delta_l = 0$, so that

$$\begin{aligned}
&\gamma_{k+1,k+1}^{lm}(\mathcal{O}_{k+1}^{rs}) \\
&= c_m(T_{k+1}) \sum_v a_{lmv} \big(\gamma_{k,k}^{mv}(\mathcal{O}_k^{rs}) + \gamma_{k,k}^{mv}(\langle X_k, e_r \rangle \langle X_{k-1}, e_s \rangle) \big) \\
&= c_m(T_{k+1}) \sum_v a_{lmv} \big(\gamma_{k,k}^{mv}(\mathcal{O}_k^{rs}) + c_m(T_{k+1}) a_{lms} \sigma_{mr} q_k(e_{ms}) \big).
\end{aligned}$$

Now take $H_{k+1} = H_k = ... = H_n = \mathcal{O}_n^{rs}$ for $k > n$, $\alpha_l = 0$, $\beta_l = 0$, $\delta_l = 0$ for $k+1 \geq l > n$. Thus we obtain the smoother

$$\gamma_{n,k+1}^{lm}(\mathcal{O}_n^{rs}) = c_m(T_{k+1}) \sum_v a_{lmv} \gamma_{n,k}^{lm}(\mathcal{O}_n^{rs}).$$

7.5.5 Estimators for state to observation transitions

We consider a process of the form

$$\mathcal{T}_k^{rs} = \sum_{l=1}^{k} \langle X_{l-1}, e_r \rangle \langle T_l, f_s \rangle,$$

where $1 \leq r \leq N$ and $1 \leq s \leq M$.
Take $H_{k+1} = \mathcal{T}_{k+1}^{rs}$, $H_0 = 0$, $\alpha_l = 0$, $\beta_l = 0$, $\delta_l = \langle X_{l-1}, e_r \rangle f_s$, then

$$\begin{aligned}
&\gamma_{k+1,k+1}^{lm}(\mathcal{T}_{k+1}^{rs}) \\
&= c_m(T_{k+1}) \sum_v a_{lmv} \big(\gamma_{k,k}^{mv}(\mathcal{T}_k^{rs}) + \langle \gamma_{k,k}^{mv}(\langle X_k, e_r \rangle f_s), T_{k+1} \rangle \big) \\
&= c_m(T_{k+1}) \sum_v a_{lmv} \big(\gamma_{k,k}^{mv}(\mathcal{T}_k^{rs}) + T_{k+1}^s \sigma_{mr} q_k(e_{mv}) \big).
\end{aligned}$$

Let $H_{k+1} = H_k = ... = H_n = \mathcal{T}_n^{rs}$ for $k+1 > n$, so that $\alpha_l = 0$, $\beta_l = 0$, $\delta_l = 0$ for $k+1 \geq l > n$. Thus

$$\gamma_{n,k+1}^{lm}(\mathcal{T}_n^{rs}) = c_m(T_{k+1}) \sum_v a_{lmv} \gamma_{n,k}^{mv}(\mathcal{T}_n^{rs}).$$

Remark 1.

$$\sum_{l,m} \gamma_{k+1,k+1}^{lm}(H_{k+1}) = \gamma_{k+1}(H_{k+1}),$$

Remark 2. Some of the recursive filters contain the term q_k, but q_k itself has the recursive filter given by (7.5). Thus we can calculate q_k recursively first, then obtain other estimates from the respective recursive filters.

7.6 Parameter re-estimations

In this section we assume that the parameters $\{a_{jik}\}$ and $\{c_{ji}\}$ are unknown. We employ the method of Expectation-Maximization (EM) algorithm and the filters developed in previous sections to estimate the parameters recursively. The Expectation-Maximization (EM) algorithm is as follows. See Baum and Petrie [3] and Dembo and Zeitouni [4]

Step 1. Set $p = 0$ and choose $\hat{\theta}_0$;
Step 2. (E-step) Set $\theta^* = \hat{\theta}_p$ and compute $Q(\cdot, \theta^*)$, where

$$Q(\theta, \theta^*) = E_{\theta^*}\left(log \frac{dP_\theta}{dP_{\theta^*}} \,\middle|\, \mathcal{Y} \right);$$

Step 3. (M-step) Find $\hat{\theta}_{p+1} \in \arg\max_{\theta \in \Theta} Q(\theta, \theta^*)$;
Step 4. Replace p by $p+1$ and repeat beginning with step 2 until a stopping criterion is satisfied.

The sequence $\{\hat{\theta}_p, p \leq 0\}$ thus generated is nondecreasing and by Jensen's inequality:

$$logL(\hat{\theta}_{p+1}) - logL(\hat{\theta}_p) \geq Q(\hat{\theta}_{p+1}, \hat{\theta}_p),$$

where $L(\theta) = E_0(\frac{dP_\theta}{dP_0}|\mathcal{Y})$ and $Q(\hat{\theta}_{p+1}, \hat{\theta}_p)$ is non-negative by the selection of $\hat{\theta}_{p+1}$. The quantities $Q(\theta, \theta^*)$ are called the conditional pseudo-log-likelihoods.

Consider the parameter space $\theta := (a_{jik}, 1 \leq j, i, k \leq N, c_{ji}, 1 \leq j \leq M, 1 \leq i \leq N)$, and the space of estimators $\hat{\theta} := (\hat{a}_{jik}, 1 \leq j, i, k \leq N, \hat{c}_{ji}, 1 \leq j \leq M, 1 \leq i \leq N)$.
In order to estimate the entries of the transition matrix A, we consider the likelihood ratio:

$$\Lambda_k^A = \prod_{l=2}^{k} \prod_{r,s,t=1}^{N} \left(\frac{\hat{a}_{trs}}{a_{trs}}\right)^{\langle X_l, e_t\rangle \langle X_{l-1}, e_r\rangle \langle X_{l-2}, e_s\rangle},$$

and set $\frac{dP_{\hat{\theta}}}{dP_\theta}\Big|_{\mathcal{F}_k} = \Lambda_k^A$. Then we have the next lemma, which shows that under $P_{\hat{\theta}}$, X has transition matrix $(\hat{a}_{trs})$.

Lemma 7. *Under the probability measure $P_{\hat{\theta}}$, assuming $X_k = e_r$ and $X_{k-1} = e_s$, then $E_{\hat{\theta}}(\langle X_{k+1}, e_t\rangle | \mathcal{F}_k) = \hat{a}_{trs}$.*

Proof.

$$\begin{aligned}
&E_{\hat{\theta}}\big(\langle X_{k+1}, e_t\rangle \,\big|\, \mathcal{F}_k\big) = \\
&= \frac{E\big(\langle X_{k+1}, e_t\rangle \Lambda_{k+1}^A \,\big|\, \mathcal{F}_k\big)}{E(\Lambda_{k+1}^A \,|\, \mathcal{F}_k)} \\
&= \frac{E\Big(\langle X_{k+1}, e_t\rangle \prod_{r',s',t'} \left(\frac{\hat{a}_{t'r's'}}{a_{t'r's'}}\right)^{\langle X_{k+1}, e_{t'}\rangle \langle X_k, e_{r'}\rangle \langle X_{k-1}, e_{s'}\rangle} \,\Big|\, \mathcal{F}_k\Big)}{E\Big(\prod_{r',s',t'} \left(\frac{\hat{a}_{t'r's'}}{a_{t'r's'}}\right)^{\langle X_{k+1}, e_{t'}\rangle \langle X_k, e_{r'}\rangle \langle X_{k-1}, e_{s'}\rangle} \,\Big|\, \mathcal{F}_k\Big)} \\
&= \frac{E\Big(\langle X_{k+1}, e_t\rangle \frac{\hat{a}_{trs}}{a_{trs}} \Big| \mathcal{F}_k\Big)}{E\Big(\sum_{t'} \langle X_{k+1}, e_{t'}\rangle \frac{\hat{a}_{t'rs}}{a_{t'rs}} \,\Big|\, \mathcal{F}_k\Big)} \\
&= \frac{\frac{\hat{a}_{trs}}{a_{trs}} E\big(\langle X_{k+1}, e_t\rangle \,\big|\, \mathcal{F}_k\big)}{\frac{\hat{a}_{t'rs}}{a_{t'rs}} E\big(\sum_{t'} \langle X_{k+1}, e_{t'}\rangle \,\big|\, \mathcal{F}_k\big)} \\
&= \frac{\frac{\hat{a}_{trs}}{a_{trs}} a_{trs}}{\sum_{t'} \frac{\hat{a}_{t'rs}}{a_{t'rs}} a_{t'rs}} \\
&= \frac{\hat{a}_{trs}}{\sum_{t'} \hat{a}_{t'rs}} \\
&= \hat{a}_{trs}.
\end{aligned}$$

Now recall that for any process ϕ_k, $k \in \mathbb{N}$, $\hat{\phi}_k = E(\phi_k | \mathcal{Y}_k)$ denotes expectation taken under the probability measure P_θ.

Theorem 5. *The new estimate of $\hat{a}_{trs}$, given the observations up to time k, is given by*

$$\hat{a}_{trs} = \frac{\hat{\mathcal{J}}_k^{trs}}{\hat{\mathcal{O}}_k^{rs}} = \frac{\gamma_k(\mathcal{J}_k^{trs})}{\gamma_k(\mathcal{O}_k^{rs})},$$

where the expectation γ_k is taken with respect to P_θ.

Proof. Consider the log-likelihood ratio:

$$\begin{aligned}\log \Lambda_k^A &= \sum_{r,s,t}\sum_{l=2}^{k} \langle X_l, e_t\rangle\langle X_{l-1}, e_r\rangle\langle X_{l-2}, e_s\rangle\big(\log \hat{a}_{trs}(k) - \log a_{trs}\big)\\ &= \sum_{r,s,t} \mathcal{J}_k^{trs} \log \hat{a}_{trs} + R(a),\end{aligned}$$

where $R(a)$ does not involve $(\hat{a}_{trs})$. Then

$$E(\log \Lambda_k^A \mid \mathcal{Y}_k) = \sum_{r,s,t} \hat{\mathcal{J}}_k^{trs} \log \hat{a}_{trs} + \hat{R}(a).$$

Next we note that $\sum_t \hat{a}_{trs} = 1$, $\sum_{t,r,s} \hat{\mathcal{O}}_k^{rs} \hat{a}_{trs} = k-1$, and $\sum_{t,r,s} \hat{\mathcal{J}}_k^{trs} = k-1$. The conditional pseudo-log-likelihood with Lagrange multiplier λ is given by

$$L^A(\hat{a}, \lambda) = \sum_{t,r,s} \hat{\mathcal{J}}_k^{rs} \log \hat{a}_{trs} + \hat{R}(a) + \lambda\Big(\sum_{t,r,s} \hat{\mathcal{O}}_k^{rs} \hat{a}_{trs} - k + 1\Big).$$

Differentiate with respect to λ and $\hat{a}_{trs}$, and then equate the derivatives to 0 to obtain

$$\frac{1}{\hat{a}_{trs}} \hat{\mathcal{J}}_k^{trs} + \lambda \hat{\mathcal{O}}_k^{rs} = 0, \quad \sum_{t,r,s} \hat{\mathcal{O}}_k^{rs} \hat{a}_{trs} = k-1$$

which give $\lambda = -1$. Thus

$$\hat{a}_{trs} = \frac{\hat{\mathcal{J}}_k^{trs}}{\hat{\mathcal{O}}_k^{rs}} = \frac{\gamma_k(\mathcal{J}_k^{trs})}{\gamma_k(\mathcal{O}_k^{rs})},$$

which maximizes the conditional pseudo-log-likelihood $L^A(\hat{a}, \lambda)$.

Next we consider the convergence property of the EM algorithm. Suppose that $\theta_p := \{\hat{a}_{trs}^p, 1 \le t, r, s \le N\}$ is a set of parameter estimates after iterating the EM algorithm p times with the given observations up to time k, where $p \ge 0$. We then have the following convergence theorem:

Theorem 6. *$L(\theta_p)$ converges monotonically to some random variable L^* in $(\Omega, \mathcal{Y}_k, P_{\theta_0})$ and any limit point θ^* of $\{\theta_p\}$ is a stationary point of $L(\theta)$, i.e. $L(\theta^*) = L^*$.*

Proof. First note that

$$L(\theta_p) = E_0\Big(\prod_{l=2}^{k} \prod_{r,s,t=1}^{N} \Big(\frac{\hat{a}_{trs}^p}{a_{trs}^0}\Big)^{\langle X_l, e_t\rangle\langle X_{l-1}, e_r\rangle\langle X_{l-2}, e_s\rangle} \,\Big|\, \mathcal{Y}_k\Big)$$

which is bounded above by

$$E_0\Big(\prod_{l=2}^{k}\prod_{r,s,t=1}^{N}(a^0_{trs})^{-\langle X_l,e_t\rangle\langle X_{l-1},e_r\rangle\langle X_{l-2},e_s\rangle}\;\Big|\;\mathcal{Y}_k\Big),$$

and $\{L(\theta_p)\}$ is increasing and converges to some random variable L^* in $(\Omega, \mathcal{Y}_k, P_{\theta_0})$. Since the function $L(\theta)$ is continous in θ, therefore any limit point θ^* of $\{\theta_p\}$ satisfies $L(\theta^*) = L^*$.

Next consider the entries c_{sr} of the state to observation transition matrix C. To replace the parameters $\theta = \{c_{sr}, 1 \le s \le M, 1 \le r \le N\}$ by $\hat{\theta} = \{\hat{c}_{sr}, 1 \le s \le M, 1 \le r \le N\}$, we perform a reverse change of measure to construct $P_{\hat{\theta}}$ by setting

$$\left.\frac{dP_{\hat{\theta}}}{d\bar{P}}\right|_{\mathcal{G}_k} = \Lambda_k^{\hat{c}},$$

where $\Lambda_k^{\hat{c}} = \prod_{l=1}^{k}\lambda_l^{\hat{c}}$, $\lambda_l^{\hat{c}} = \prod_{s=1}^{M}(M\hat{c}_l^s)^{T_l^s}$ with $l,k \in \mathbb{N}$, $\hat{c}_l^s = \langle \hat{C}X_{l-1}, f_s\rangle$, and $\hat{C} = (\hat{c}_{sr})$ is a state to observation transition matrix. Similar to Lemma 6 we have the following.

Lemma 8. *Under the probability measure $P_{\hat{\theta}}$, $E_{\hat{\theta}}(T_{k+1}|\mathcal{G}_k) = \hat{C}X_k$.*

It shows that under $P_{\hat{\theta}}$, the state to observation transition matrix is $\hat{C}$. Notice that $\hat{c}_l^s = \sum_{r=1}^{N}\hat{c}_{sr}\langle X_{l-1}, e_r\rangle = \prod_{r=1}^{N}\hat{c}_{sr}^{\langle X_{l-1},e_r\rangle}$, hence the likelihood function has the form:

$$\begin{aligned}
\Lambda_k^C :&= \left.\frac{dP_{\hat{\theta}}}{dP}\right|_{\mathcal{G}_k}\\
&= \left.\frac{dP_{\hat{\theta}}}{d\bar{P}}\frac{d\bar{P}}{dP}\right|_{\mathcal{G}_k}\\
&= \Lambda_k^{\hat{c}}\Lambda_k\\
&= \prod_{l=1}^{k}\prod_{r=1}^{N}\prod_{s=1}^{M}\Big(\frac{\hat{c}_{sr}}{c_{sr}}\Big)^{\langle X_{l-1},e_r\rangle\langle T_l,f_s\rangle}.
\end{aligned}$$

Since

$$\sum_{s=1}^{M}\hat{c}_{sr} = 1;\ \sum_{l=1}^{k}\sum_{r=1}^{N}\sum_{s=1}^{M}\langle X_k, e_r\rangle\hat{c}_{sr} = k;\ \sum_{r=1}^{N}\sum_{s=1}^{M}\hat{\mathcal{O}}_k^r\hat{c}_{sr} = k$$

and

$$\sum_{r=1}^{N}\sum_{s=1}^{M}\mathcal{T}_k^{rs} = k,$$

thus

$$E(\log \Lambda_k^C \mid \mathcal{Y}_k) = \sum_{r=1}^{N}\sum_{s=1}^{M} \hat{\mathcal{T}}_k^{rs} \log \hat{c}_{sr} + \hat{R}(c),$$

where $\hat{R}(c)$ does not involve $(\hat{c}_{sr})$. With the Lagrange multiplier λ, we obtain the conditional pseudo-log-likelihood:

$$L^C(\hat{c}, \lambda) = \sum_{r=1}^{N}\sum_{s=1}^{M} \hat{\mathcal{T}}_k^{rs} log \hat{c}_{sr} + \hat{R}(c) + \lambda\Big(\sum_{r=1}^{N}\sum_{s=1}^{M} \hat{\mathcal{O}}_k^r \hat{c}_{sr} - k\Big).$$

Differentiate with respect to λ and $\hat{c}_{sr}$, and equate the derivatives to 0, we obtain the maximum conditional pseudo-log-likelihood estimates $\hat{c}_{sr}$ given as

$$\hat{c}_{sr} = \frac{\hat{\mathcal{T}}_k^{rs}}{\hat{\mathcal{O}}_k^r} = \frac{\gamma_k(\mathcal{T}_k^{rs})}{\gamma_k(\mathcal{O}_k^r)}.$$

If $\theta_p := \{\hat{c}_{rs}^p, 1 \le r \le M, 1 \le s \le N\}$ denotes the set of parameter estimates after iterating the EM algorithm p times with the given observations up to time k, where $p \ge 0$, we obtain the following convergence theorem:

Theorem 7. *$L(\theta_p)$ converges monotonely to some r.v. L^* in $(\Omega, \mathcal{Y}_k, P_{\theta_0})$ and any limit point θ^* of $\{\theta_p\}$ is a stationary point of $L(\theta)$, i.e. $L(\theta^*) = L^*$.*

7.7 Error analysis

In this section we give recursive estimates of the conditional mean square errors resulted from our filters obtained in the previous sections. First note that

$$E\big[(H_{k+1} - \hat{H}_{k+1})^2 \mid \mathcal{Y}_{k+1}\big] = E\big(H_{k+1}^2 \mid \mathcal{Y}_{k+1}\big) - \hat{H}_{k+1}^2,$$

Thus it remains to compute $E(H_{k+1}^2|\mathcal{Y}_{k+1})$ and hence obtain the conditional variance.

Consider H_{k+1} when $\beta_{k+1} = 0$, so that $H_{k+1} = H_k + \alpha_{k+1} + \langle \delta_{k+1}, V_{k+1}\rangle$, and

$$\begin{aligned} H_{k+1}^2 = {} & H_k^2 + (2H_k + \alpha_{k+1})\alpha_{k+1} + \langle 2(H_k + \alpha_{k+1})\delta_{k+1}, T_{k+1}\rangle \\ & + \langle \delta_{k+1}, T_{k+1}\rangle^2. \end{aligned}$$

By a proof similar to that of Theorem 4, we obtain

$$\begin{aligned} \gamma_{k+1,k+1}^{ls}(\langle \delta_{k+1}, T_{k+1}\rangle^2) &= \gamma_{k+1,k+1}^{ls}(T_{k+1}'\delta_{k+1}\delta_{k+1}'T_{k+1}) \\ &= T_{k+1}'\gamma_{k+1,k+1}^{ls}(\delta_{k+1}\delta_{k+1}')T_{k+1}. \end{aligned}$$

Thus we have the recursive formula

$$\begin{aligned}\gamma^{ls}_{k+1,k+1}(H^2_{k+1}) =& c_s(T_{k+1})\sum_v a_{lsv}\big[\gamma^{sv}_{k,k}(H^2_k+\alpha_{k+1}(2H_k+\alpha_{k+1}))\\ &+\langle\gamma^{s,v}_{k,k}(2(H_k+\alpha_{k+1})\delta_{k+1}),T_{k+1}\rangle\\ &+T'_{k+1}\gamma^{ls}_{k+1,k+1}(\delta_{k+1}\delta'_{k+1})T_{k+1}\big].\end{aligned} \tag{7.6}$$

Finally we consider the conditional covariance matrix of the error resulting from our state estimates. Denote the conditional covariance matrix of X_k given $\mathcal{Y}_k$ by Σ_k and the ij^{th} element of Σ_k by σ^k_{ij}, i.e. $\Sigma_k = (\sigma^k_{ij})$, where $1 \leq i,j \leq N$. Since $X_k = \sum_{r=1}^{N}\langle X_k, e_r\rangle e_r$, we have

$$\begin{aligned}\sigma^k_{ij} &= E\big[(\langle X_k,e_i\rangle - E(\langle X_k,e_i\rangle \mid \mathcal{Y}_k))(\langle X_k,e_j\rangle - E(\langle X_k,e_j\rangle \mid \mathcal{Y}_k) \bigm| \mathcal{Y}_k\big]\\ &= E\big(\langle X_k,e_i\rangle\langle X_k,e_j\rangle \bigm| \mathcal{Y}_k\big) - E\big(\langle X_k,e_i\rangle \bigm| \mathcal{Y}_k\big)E\big(\langle X_k,e_j\rangle \bigm| \mathcal{Y}_k\big),\end{aligned}$$

Hence

$$\sigma^k_{ij} = \begin{cases} p_k(e_i) - p^2_k(e_i) & if \;\; i=j \\ -p_k(e_i)p_k(e_j) & if \;\; i \neq j \end{cases}.$$

Thus we can calculate the conditional covariance recursively since the p_k's can be computed recursively.

7.8 Conclusion

In this paper, we studied a hidden weak Markov model with discrete time, finite state and observation space. We developed a general recursive filter which covers the case of state, occupation time, and total number of jumps from one state to another state filter estimates. We then provided an EM algorithm to re-estimate the parameters of our model. For simplicity and without loss of generality, we considered the case of memory order 2. For the general case, when we have a weak Markov chain of order $k > 2$, one way of expressing the k-step transition matrix A would be

$$\begin{pmatrix} a_{11\ldots11} & a_{11..12} & \cdots & a_{11\ldots1N} & \cdots & a_{1N1\ldots11} & \cdots & a_{1N\ldots NN} \\ a_{21\ldots11} & a_{21..12} & \cdots & a_{21\ldots1N} & \cdots & a_{2N1\ldots11} & \cdots & a_{2N\ldots NN} \\ & & & \cdots & & \cdots & & \\ a_{N1\ldots11} & a_{N1..12} & \cdots & a_{N1\ldots1N} & \cdots & a_{NN1\ldots11} & \cdots & a_{NN\ldots NN} \end{pmatrix}$$

which is an $N \times N^k$ matrix. The sub-index of $a_{(\cdot)}$ has the form $i_1j_1j_2...j_k$ where i_1 corresponds to the row number and $j_1j_2...j_k$ is a string of k numbers in $\{1,2,...,N\}$ which corresponds to the column numbers. As a consequence, all the equations will be adjusted accordingly. For example, when $k=3$, the transition matrix A is given by

$$\begin{pmatrix} a_{1111} & a_{1112} & \dots & a_{111N} & \dots & a_{1N11} & \dots & a_{1NNN} \\ a_{2111} & a_{2112} & \dots & a_{211N} & \dots & a_{2N11} & \dots & a_{2NNN} \\ & & & \dots & & \dots & & \\ a_{N111} & a_{N112} & \dots & a_{N11N} & \dots & a_{NN11} & \dots & a_{NNNN} \end{pmatrix}$$

which is a $N \times N^3$ matrix, and part (iii) of Proposition 1 becomes

$$E(X_{k+1} \mid \mathcal{G}_k) = A\alpha \begin{pmatrix} X_k \\ X_{k-1} \\ X_{k-2} \end{pmatrix} = E(X_{k+1} \mid X_k, X_{k-1}, X_{k-2})$$

where $\alpha \begin{pmatrix} e_r \\ e_s \\ e_t \end{pmatrix} = (0...0 \;\; 1 \;\; 0...0)' \in \mathbb{R}^{N^3}$, for $1 \le r, s, t \le N$, where 1 is at $((r-1)N^2 + (s-1)N + t)^{th}$ position.

References

1. Beran, J. (1994). *Statistics For Long-Memory Processes*, Chapman and Hall Publishing.
2. Cutland, N. J., Kopp, P. E. and W. Willinger (1995). "Stock price returns and the Joseph effect, a fractional version of the Black-Sholes model,seminar on stochastic analysis, random fields and applications". *Prog. Probability, Ascona* 36 (1993) *Birkhauser*, 327–351.
3. Baum, L.E. and T. Petrie (1966). "Statistical inference for probabilistic functions of finite state Markov chains". *Annals of the Institute of Statistical Mathematics*, 37: 1554–1563.
4. Dembo, A. and O. Zeitouni (1986). "Parameter estimation of partially observed continuous time stochastic processes via the EM algorithm". *Stochastic Processes and their Applications*, 23: 91–113.
5. Elliott, R.J., Aggoun, L. and J.B. Moore (1995). *Hidden Markov Models: Estimation and Control.* Springer Verlag New York (1995).
6. Elliott, R.J., Malcolm, W.P. and A.H. Tsoi (2003). "Robust parameter estimation for asset price models with Markov modulated volatilities". *Journal of Economic Dynamics and Control,* 27(8): 1391–1409.
7. Elliott, R.J. and R.W. Rishel (1982). "Estimating the implicit interest rate of a risky asset", *Stochastic Processes and their Applications*, 49: 199–206.
8. Elliott, R.J. and J. van der Hoek (2003). "A general fractional white noise theory and applications to finance". *Mathematical Finance,* 13(2): 301–330.
9. Elliott, R.J. (1994). "Exact adaptive filters for Markov chains observed in Gaussian noise". *Automatica,* 30(9): 1399–1408.
10. Elliott, R.J. (1993). "New finite dimensional filters and smoothers for noisily observed Markov chains". *IEEE Transactions on Information Theory,* 39(1): 265–271.
11. Ephraim, Y. and N. Merhav (2002). Hidden Markov processes. *IEEE Transactions on Information Theory,* 48(6): 1518–1569.

12. Frey, R. and W. Runggaldier (2001). "A nonlinear filtering approach to volatility estimation with a view towards high frequency data". *International Journal of Theoretical and Applied Finance*, 4(2): 199–210.
13. Hu, Y. and B. Oksendal (2003). "Fractional white noise calculus and applications to finance". *Infinite Dimensional Analysis, Quantum Probability and Related Topics*, 6(1): 1–32.
14. Luo, S. and A.H. Tsoi (2006). "Filtering of hidden weak Markov chains - continuous range observations", *Submitted.*
15. Luo, S., Tsoi, A.H. and P. Yin (2006). "Weak Markov modulated drift and volatility estimation", *Submitted.*
16. Luo, S. "Hidden weak Markov modeling on asset allocation", *Preprint.*
17. Mandelbrot, B.B. and J.W. van Ness (1968). "Fractional Brownian motions, fractional noises and applications". *SIAM Rev.* 10: 422–437.
18. Tsoi, A.H. "Discrete time weak Markov control - dynamic programming equation with finite horizon", *Submitted.*
19. Tsoi, A.H. "A discrete time weak Markov term structure model", *Working Paper.*
20. Tsoi, A.H. "Discrete time reversal and duality of weak Markov chain", *Working Paper.*
21. Tsoi, A.H., Yang, H. and S.N. Yeung (2000). "European option pricing when the riskfree interest rate follows a jump process". *Stochastic Models*, 16(1): 143–166.
22. Y.H. Wang (1992). "Approximation kth-order two-state markov chains". *Journal of Applied Probability,* 29: 861–868.

8

Filtering of a Partially Observed Inventory System

Lakhdar Aggoun

Department of Mathematics and Statistics
Sultan Qaboos University
P.O.Box 36, Al-Khod 123
Sultanate of Oman
laggoun@squ.edu.om

Summary. The vast majority of work done on inventory system is based on the critical assumption of fully observed inventory level dynamics and demands. Modern technology, like the internet, offers a tremendous number of opportunities to businesses to collect imperfect but useful information on potential customers which helps them planning efficiently to meet future demands. For instance visits to commercial web sites provides the management of a business of a source of partial information on future demands. On the other hand it is often the case that it is not economically viable to fully observe the dynamics of inventory levels and only partial information is accessible to the management. In this article, using hidden Markov model techniques we estimate the inventory level as well as future demands of partially observed inventory system. The parameters of the model are updated via the EM algorithm.

Key words: Filtering, Markov chains, change of measure, inventory model.

8.1 Introduction

Modern technology, like the internet, offers a tremendous number of opportunities to businesses to collect useful information which helps them planning efficiently to meet future demands. Visits to commercial web sites constitute a source of partial information on future demands of the commodities (or services) offered by companies. Warnings by e-mail (or by some other means such as mobile phone short message service) of customers on change in the price of a commodity provide a source of potential sales.

Another way of acquiring partial information on future demands is provided by a company that uses sales representatives to market its products. Each contact of a sales representative with a customer yields a potential demand. Sometimes sales representatives prepare sales vouchers as means for quoting the customers showing willingness to buy. Since it usually takes some time for a potential sale to be materialized, the collection of sales representatives' information as to the number of customers interested in a product (such as the number of outstanding sales vouchers) can generate an indication about the future sales of that product [21].

Treharne and Sox [22] discuss a non-stationary demand situation where the demand is partially observed. They model the demand as a composite-state, partially observed Markov decision process.

Another example is provided by DeCroix and Mookerjee [14] who consider a periodic review problem in which there is an option of purchasing demand information at the beginning of each period. They consider two levels of demand information: Perfect information allows the decision maker to know the exact demand of the coming period, whereas the imperfect one identifies a particular posterior demand distribution.

Karaesmen, Buzacott, and Dallery [17] consider a capacitated problem under partial information on demand and stochastic lead times. They model the problem via a discrete time make-to-stock queue.

Many factors contribute to make inventories hard to be fully observed by the management. Among these factors are thefts, shoplifting, damaged or misplaced items, low production yield processes [23], perished items [20] etc.

An earlier literature review on partially observed systems can be found in Monahan [19]. Since then, there have not been much research activities in the study of partially observed inventories.

Bensousan et al. [8, 9, 10, 11, 12], Treharne and Sox [22] study partially observed demands in the context of discrete time optimal control. In their studies, the demand is Markov-modulated but the underlying demand state is unobserved. Another example of a Markov modulated model is discussed in Beyer et al. [13].

Models discussing filtering and parameter estimation are considered by Aggoun et al. [1, 2, 3, 4].

In this article we consider a discrete-time, discrete state inventory model where the demand is a partially observed finite state process modulated by a Markov chain which is part of the information available at the beginning of each period. These two processes, in turn modulate a replenishment process. In other words the amount to be ordered and stocked to satisfy the (estimated) demand which must be met, say in the next period, relies on the partial information on futures sales collected and made available in the current period. For the sake of simplicity and to be dealing with only finite state processes, we assume that information does not accumulate without bound. That is, information on potential sales from earlier periods is discarded.

The paper's objective is to estimate recursively the joint distribution of the level of stock and the actual demand as well as to re-estimate the model parameters. This article is divided into six sections and is organized as follows. In §2 we define the model. In §3 we describe the reference probability method used in computing our filters. In §4 and §5 we derive filters for various quantities of interest. The parameters of the model are re-estimated via the EM algorithm in §6.

8.2 Model description

All processes are initially defined on a probability space $(\Omega, \mathcal{G}, P)$. Our model consists of the following components.

1. The observed number of potential demands available at the beginning of period n is an L-state discrete-time Markov process $Y = \{Y_k, 1 \leq k\}$. We use the canonical representation of a Markov chain (see [5, 16]). So, without loss of generality we take the state space for Y to be the set $\mathcal{L} = \{e_1, e_2, \ldots, e_L\}$, whose elements e_i are column vectors with unity in the ith position and zero elsewhere. The essential point of this canonical representation of a Markov chain, is that the state dynamics can be written down in the form
$$Y_n = AY_{n-1} + V_n. \tag{8.1}$$
Here V is a $(P, \sigma\{Y_1, Y_2, \ldots, Y_n\})$-martingale increment and $A \in \mathbb{R}^{L\times L}$ is a matrix of state transition probabilities such that $P(Y_n = j \mid Y_{n-1} = i) \triangleq a_{ji}$.
2. The actual (unobserved) demand process D is a finite-state process with N states $\{d_1, \ldots, d_N\}$. Without loss of generality, we identify the state space $\{d_1, \ldots, d_N\}$ with the sets of standard unit vectors $\{f_1, f_2, \ldots, f_N\}$ of $\mathbb{R}^N$. We shall assume that
$$P(D_n = f_m \mid D_1, \ldots, D_{n-1}, Y_0, Y_1, \ldots, Y_{n-1}) = P(D_n = f_m \mid D_{n-1}, Y_{n-1}).$$
Write
$$b_{m\ell i} = P(D_n = f_m \mid D_{n-1} = f_\ell, Y_{n-1} = e_i)$$
and $B = \{b_{m\ell i}\}$, $m, \ell = 1, \ldots, N;\ i = 1, \ldots, L$. Therefore $\sum_{m=1}^{N} b_{m\ell i} = 1$ and we have the semimartingale representation
$$D_n = BD_{n-1} \otimes Y_{n-1} + W_n. \tag{8.2}$$
Here W_n is a sequence of martingale increments. For (column) vectors $x \in \mathbb{R}^L$, $y \in \mathbb{R}^N$ their tensor or Kronecker product $x \otimes y$ is the vector $xy' \in \mathbb{R}^{LN}$.

3. Let the demand which was met at the beginning of the n-th period (or by the end of the $(n-1)$-th period) be denoted by $\mathfrak{D}_{n-1}$. We assume here that for $n \geq 1$, $\mathfrak{D}_n$ is a Poisson random variable with mean $\lambda_n(X_{n-1}, Y_{n-1})$. Here X_n is the inventory level at the beginning of the n-th period (see the dynamics in (8.4).
4. A replenishment process U such that for $n \geq 1$, U_n is a nonnegative integer-valued random variable with finite-support probability distribution:
$P(U_n = u \mid \mathfrak{D}_k, U_k, Y_k, D_k, X_k,\ k \leq n-1) = \phi_n(u, X_{n-1}, Y_{n-1}, D_{n-1})$.
5. Each item in the stock at the beginning of the n-th period is assumed to be perished (damaged, stolen etc.) with probability $(1-\alpha)$ independently of the other items, where $0 < \alpha < 1$ or is intact with probability α.

   We shall be using the Binomial thining operator " $\circ$ " which is well-known in Time Series Analysis [6], [18]. This operator is defined as follows.

   For any nonnegative integer-valued random variable X and $\alpha \in (0,1)$,

$$\alpha \circ X = \sum_{j=1}^{X} Y_j, \tag{8.3}$$

   where $Y_1, Y_2, \ldots$ is a sequence of independent, identically distributed (IID) random variables independent of X, such that $P(Y_j = 1) = 1 - P(Y_j = 0) = \alpha$. Now let X_n be an integer-valued random variable representing the number of items in stock at the beginning of period n in the inventory with dynamics

$$X_n = \alpha \circ X_{n-1} + U_{n-1} - \mathfrak{D}_{n-1}, \tag{8.4}$$

   with X_0 constant (integer) or its distribution known. If X_{n-1} is negative then $\alpha \circ X_{n-1} = 0$. Note that a negative X_n is interpreted as shortage.

Write the following complete filtration $\mathcal{Y}_n = \sigma\{\mathfrak{D}_k, U_k, Y_k, k \leq n\}, \mathcal{G}_n = \sigma\{\mathfrak{D}_k, U_k, Y_k, D_k, X_k,\ k \leq n\}$.

8.3 Reference probability

In our context, the objective of the method of reference probability is to choose a measure $\overline{P}$, on the measurable space $(\Omega, \mathcal{F})$, under which

(i) Process D is a sequence of IID random variables uniformly distributed on the set $\{f_1, f_2, \ldots, f_N\}$, that is $P(D_n \mid \mathcal{G}_{n-1}) = \frac{1}{N}$.
(ii) Process Y is a sequence of IID random variables uniformly distributed on the set $\{e_1, e_2, \ldots, e_L\}$, that is $P(Y_n = e_j \mid \mathcal{G}_{n-1}) = \frac{1}{L}$.
(iii) Process U is a sequence of IID random variables with some suitable positive distribution ψ.

Further, under the measure $\overline{P}$, the dynamics for X are unchanged.

The probability measure P is referred to as the 'real world' measure, that is, under this measure

$$P \quad \begin{cases} Y_n = AY_{n-1} + V_n, \\ D_n = BD_{n-1} \otimes Y_{n-1} + W_n, \\ P(\mathfrak{D}_n = \mathfrak{d} \mid \mathcal{G}_{n-1}) = \frac{1}{\mathfrak{d}!}\lambda_n^{\mathfrak{d}} e^{-\lambda_n}, \\ \text{Process } U \text{ is such that for } n \geq 1,\ U_n \text{ is a nonnegative integer-valued} \\ \text{random variable with finite-support distribution such that} \\ P(U_n = u \mid \mathcal{G}_{n-1}) = \phi_n(u, X_{n-1}, Y_{n-1}, D_{n-1}). \end{cases} \tag{8.5}$$

Definition 1. *Denote by $\Gamma = \{\gamma_k, 0 \leq k\}$ the stochastic process whose value at k is given by*

$$\Gamma_n = \prod_{k=0}^{n} \gamma_k, \tag{8.6}$$

where $\gamma_0 = 1$ and

$$\begin{aligned} \gamma_k = & \prod_{i=1}^{L} \Big(\lambda_k^{\mathfrak{D}_k}(X_{k-1}, i) e^{-\lambda_k(X_{k-1}, i)+1}\Big)^{\langle Y_{k-1}, e_i\rangle} \prod_{i,j=1}^{L} (La_{ji})^{\langle Y_k, e_j\rangle\langle Y_{k-1}, e_i\rangle} \\ & \prod_{m,\ell=1}^{N} \prod_{i=1}^{L} (Nb_{m\ell i})^{\langle D_k, f_m\rangle\langle D_{k-1}, f_\ell\rangle\langle Y_{k-1}, e_i\rangle} \\ & \prod_{\ell=1}^{N} \prod_{i=1}^{L} \Big(\frac{\phi_k(U_k, X_{k-1}, i, \ell)}{\psi(U_k)}\Big)^{\langle Y_{k-1}, e_i\rangle\langle D_{k-1}, f_\ell\rangle}. \end{aligned} \tag{8.7}$$

We define the 'real world' measure P in terms of $\overline{P}$, by setting $\left.\frac{dP}{d\overline{P}}\right|_{\mathcal{G}_n} \triangleq \Gamma_n$. The existence of P follows from the Kolmogorov Extension Theorem.

8.4 Filtering

Write

$$E\Big[\langle D_n, f_v\rangle I(X_n = x) \,\Big|\, \mathcal{Y}_n\Big] = \frac{\overline{E}\Big[\Gamma_n \langle D_n, f_v\rangle I(X_n = x) \,\Big|\, \mathcal{Y}_n\Big]}{\overline{E}\Big[\Gamma_n \,\Big|\, \mathcal{Y}_n\Big]}$$

and

$$\overline{E}\Big[\Gamma_n \langle D_n, f_v\rangle I(X_n = x) \,\Big|\, \mathcal{Y}_n\Big] = \rho_n(v, x).$$

Theorem 1. *Denote by* $\rho_0(v,x)$*, the initial probability distribution of* (D_0, X_0)*. The unnormalised probability* $\rho_n(v,x)$*, satisfies the recursion*

$$\rho_n(v,x) = \prod_{i,j=1}^{L} (La_{ji})^{\langle Y_n, e_j\rangle\langle Y_{n-1}, e_i\rangle} \sum_{\ell=1}^{N}\sum_{i=1}^{L} b_{v\ell i}\langle Y_{n-1}, e_i\rangle \sum_{\mathfrak{z} \geq x - U_{n-1} + \mathfrak{D}_{n-1}}$$

$$\mathrm{Bin}(\mathfrak{z}, \alpha, x - U_{n-1} + \mathfrak{D}_{n-1})\lambda_n^{\mathfrak{D}_n}(\mathfrak{z}, i, \ell)\mathrm{e}^{-\lambda_n(\mathfrak{z},i)+1}\frac{\phi_k(U_n, \mathfrak{z}, i, \ell)}{\psi(U_n)}\rho_{n-1}(\ell, \mathfrak{z}),$$

where

$$\mathrm{Bin}(\mathfrak{z}, \alpha, x - U_{n-1} + \mathfrak{D}_{n-1}) = \binom{\mathfrak{z}}{x - U_{n-1} + \mathfrak{D}_{n-1}} (\alpha)^{x - U_{n-1} + \mathfrak{D}_{n-1}}(1-\alpha)^{\mathfrak{z} - x + U_{n-1} - \mathfrak{D}_{n-1}}.$$

Proof. In view of (8.7), (8.6) and the independence assumptions under $\overline{P}$

$$\begin{aligned}
\rho_n(v,x) &= \prod_{i,j=1}^{L} (La_{ji})^{\langle Y_n,e_j\rangle\langle Y_{n-1},e_i\rangle} \overline{E}\Big[\Gamma_{n-1}\langle D_n, f_v\rangle \\
&\times \prod_{m,\ell=1}^{N}\prod_{i=1}^{L}(Nb_{m\ell i})^{\langle D_n,f_m\rangle\langle D_{n-1},f_\ell\rangle\langle Y_{n-1},e_i\rangle}\lambda_n^{\mathfrak{D}_n}(X_{n-1},i)\mathrm{e}^{-\lambda_n(X_{n-1},i)+1} \\
&\times I(X_n = x)\prod_{\ell=1}^{N}\prod_{i=1}^{L}\Big(\frac{\phi_n(U_n,X_{n-1},i,\ell)}{\psi(U_n)}\Big)^{\langle Y_{n-1},e_i\rangle\langle D_{n-1},f_\ell\rangle}\ \Big|\ \mathcal{Y}_n\Big] \\
&= \prod_{i,j=1}^{L} (La_{ji})^{\langle Y_n,e_j\rangle\langle Y_{n-1},e_i\rangle}\sum_{\ell=1}^{N}\sum_{i=1}^{L} b_{v\ell i}\langle Y_{n-1},e_i\rangle\frac{1}{\psi(U_n)} \\
&\times \overline{E}\Big[\Gamma_{n-1}\phi_k(U_n,X_{n-1},i,\ell)\langle D_{n-1},f_\ell\rangle\lambda_n^{\mathfrak{D}_n}(X_{n-1},i)\mathrm{e}^{-\lambda_n(X_{n-1},i)+1} \\
&\times I(\alpha\circ X_{n-1}+U_{n-1}-\mathfrak{D}_{n-1}=x)\ \Big|\ \mathcal{Y}_n\Big] \\
&= \prod_{i,j=1}^{L} (La_{ji})^{\langle Y_n,e_j\rangle\langle Y_{n-1},e_i\rangle}\sum_{\ell=1}^{N}\sum_{i=1}^{L} b_{v\ell i}\langle Y_{n-1},e_i\rangle\frac{1}{\psi(U_n)} \\
&\times \overline{E}\Big[\Gamma_{n-1}\phi_k(U_n,X_{n-1},i,\ell)\langle D_{n-1},f_\ell\rangle\lambda_n^{\mathfrak{D}_n}(X_{n-1},i)\mathrm{e}^{-\lambda_n(X_{n-1},i)+1} \\
&\times I(\alpha\circ X_{n-1}=x-U_{n-1}+\mathfrak{D}_{n-1})\ \Big|\ \mathcal{Y}_{n-1}\Big] \\
&= \prod_{i,j=1}^{L} (La_{ji})^{\langle Y_n,e_j\rangle\langle Y_{n-1},e_i\rangle}\sum_{\ell=1}^{N}\sum_{i=1}^{L} b_{v\ell i}\langle Y_{n-1},e_i\rangle\sum_{\mathfrak{z}\geq x-U_{n-1}+\mathfrak{D}_{n-1}} \\
&\times \mathrm{Bin}(\mathfrak{z},\alpha,x-U_{n-1}+\mathfrak{D}_{n-1})\lambda_n^{\mathfrak{D}_n}(\mathfrak{z},i)\mathrm{e}^{-\lambda_n(\mathfrak{z},i,\ell)+1} \\
&\times \frac{\phi_k(U_n,\mathfrak{z},i,\ell)}{\psi(U_n)}\overline{E}\Big[\Gamma_{n-1}\langle D_{n-1},f_\ell\rangle I(X_{n-1}=\mathfrak{z})\ \Big|\ \mathcal{Y}_{n-1}\Big] \\
&= \prod_{i,j=1}^{L} (La_{ji})^{\langle Y_n,e_j\rangle\langle Y_{n-1},e_i\rangle}\sum_{\ell=1}^{N}\sum_{i=1}^{L} b_{v\ell i}\langle Y_{n-1},e_i\rangle\sum_{\mathfrak{z}\geq x-U_{n-1}+\mathfrak{D}_{n-1}} \\
&\times \mathrm{Bin}(\mathfrak{z},\alpha,x-U_{n-1}+\mathfrak{D}_{n-1})\lambda_n^{\mathfrak{D}_n}(\mathfrak{z},i,\ell)\mathrm{e}^{-\lambda_n(\mathfrak{z},i)+1} \\
&\times \frac{\phi_k(U_n,\mathfrak{z},i,\ell)}{\psi(U_n)}\rho_{n-1}(\ell,\mathfrak{z}).
\end{aligned}$$

Here

$$\begin{aligned}
&\mathrm{Bin}(\mathfrak{z},\alpha,x-U_{n-1}+\mathfrak{D}_{n-1}) \\
&= \binom{\mathfrak{z}}{x-U_{n-1}+\mathfrak{D}_{n-1}}(\alpha)^{x-U_{n-1}+\mathfrak{D}_{n-1}}(1-\alpha)^{\mathfrak{z}-x+U_{n-1}-\mathfrak{D}_{n-1}}.
\end{aligned}$$

Remark 1. The method we develop to estimate the parameters of the model, is based upon the filter for process D and estimating a set of quantities derived from it. These quantities and others of interest are listed below.

1. $\mathfrak{T}_n^{(j,i)}$, a discrete time counting process for the transitions $e_i \to e_j$ of the (observed) Markov chain Y, where $i \neq j$,

$$\mathfrak{T}_n^{(j,i)} = \sum_{k=1}^{n} \langle Y_{k-1}, e_i \rangle \langle Y_k, e_j \rangle. \tag{8.8}$$

2. $\mathfrak{J}_n^i$, the cumulative sojourn time spent by the Markov chain Y in state e_i,

$$\mathfrak{J}_n^i = \sum_{k=1}^{n} \langle Y_{k-1}^k, e_i \rangle. \tag{8.9}$$

3. $G_n^{m\ell i}$, the number of times the process D jumps from state f_ℓ to state f_m while the Markov chain Y is in state e_i.

$$G_n^{m\ell i} = \sum_{k=1}^{n} \langle D_{k-1}, f_\ell \rangle \langle D_k, f_m \rangle \langle Y_{k-1}, e_i \rangle. \tag{8.10}$$

4. $S_n^{\ell i}$, the number of times the process D is in state f_ℓ while the Markov chain Y is in state e_i.

$$S_n^{\ell i} = \sum_{k=1}^{n} \langle D_{k-1}, f_\ell \rangle \langle Y_{k-1}, e_i \rangle. \tag{8.11}$$

8.5 Filters for $G_n^{m\ell i}$, and $S_n^{\ell i}$

Rather than directly estimating the quantities, $G_n^{m\ell i}$, and $S_n^{\ell i}$ recursive forms can be found by estimating the related product-quantities, $G_n^{m\ell i} D_n I(X_n = x) \in \mathbb{R}^N$, etc. The outputs of these filters can then be manipulated to marginalise out the processes X and D, resulting in filtered estimates of the quantities of primary interest.
Write

$$\mathfrak{q}_n(G_n^{m\ell i} D_n I(X_n = x)) \triangleq \overline{E}\big[\Gamma_n G_n^{m\ell i} D_n I(X_n = x) \mid \mathcal{Y}_n\big] \quad \text{and}$$

$$\mathfrak{q}_n(S_n^{\ell i} D_n I(X_n = x)) \triangleq \overline{E}\big[\Gamma_n S_n^{\ell i} D_n \mid \mathcal{Y}_n\big].$$

Theorem 2. *The processes defined above are computed recursively by the dynamics*

$$\begin{aligned}
&\mathfrak{q}_n(G_n^{m\ell i} D_n I(X_n = x)) \\
&= \prod_{i,j=1}^{L} (La_{ji})^{\langle Y_n, e_j\rangle\langle Y_{n-1}, e_i\rangle} \\
&\quad \times \sum_{l,t=1}^{N} \sum_{i=1}^{L} b_{tli} \langle Y_{n-1}, e_i\rangle \sum_{\mathfrak{z} \geq x - U_{n-1} + \mathfrak{D}_{n-1}} \mathrm{Bin}(\mathfrak{z}, \alpha, x - U_{n-1} + \mathfrak{D}_{n-1}) \\
&\quad \times \lambda_n^{\mathfrak{D}_n}(\mathfrak{z}, i, l) \mathrm{e}^{-\lambda_n(\mathfrak{z}, i, l)+1} \frac{\phi_n(U_n, \mathfrak{z}, i, l)}{\psi(U_n)} \langle \mathfrak{q}_{n-1}(G_{n-1}^{m\ell i} D_{n-1} I(X_{n-1} = \mathfrak{z})), f_l \rangle f_t \\
&\quad + \prod_{j=1}^{L} (La_{ji})^{\langle Y_n, e_j\rangle\langle Y_{n-1}, e_i\rangle} \sum_{i=1}^{L} b_{m\ell i} \langle Y_{n-1}, e_i\rangle \sum_{\mathfrak{z} \geq x - U_{n-1} + \mathfrak{D}_{n-1}} \mathrm{Bin}(\mathfrak{z}, \alpha, x - U_{n-1} + \mathfrak{D}_{n-1}) \\
&\quad \times \lambda_n^{\mathfrak{D}_n}(\mathfrak{z}, i, \ell) \mathrm{e}^{-\lambda_n(\mathfrak{z}, i, \ell)+1} \frac{\phi_n(U_n, \mathfrak{z}, i, \ell)}{\psi(U_n)} \rho_{n-1}(\ell, \mathfrak{z}) f_m.
\end{aligned}$$

$$\begin{aligned}
\mathfrak{q}_n(S_n^{\ell i} D_n I(X_n = x)) &= \prod_{i,j=1}^{L} (La_{ji})^{\langle Y_n, e_j\rangle\langle Y_{n-1}, e_i\rangle} \\
&\quad \times \sum_{l,t=1}^{N} \sum_{i=1}^{L} b_{tli} \langle Y_{n-1}, e_i\rangle \sum_{\mathfrak{z} \geq x - U_{n-1} + \mathfrak{D}_{n-1}} \mathrm{Bin}(\mathfrak{z}, \alpha, x - U_{n-1} + \mathfrak{D}_{n-1}) \\
&\quad \times \lambda_n^{\mathfrak{D}_n}(\mathfrak{z}, i, l) \mathrm{e}^{-\lambda_n(\mathfrak{z}, i, l)+1} \frac{\phi_n(U_n, \mathfrak{z}, i, l)}{\psi(U_n)} \\
&\quad \times \langle \mathfrak{q}_{n-1}(S_{n-1}^{\ell i} D_{n-1} I(X_{n-1} = \mathfrak{z})), f_l \rangle f_t + \prod_{j=1}^{L} (La_{ji})^{\langle Y_n, e_j\rangle\langle Y_{n-1}, e_i\rangle} \\
&\quad \times \sum_{t=1}^{N} \sum_{i=1}^{L} b_{t\ell i} \langle Y_{n-1}, e_i\rangle \sum_{\mathfrak{z} \geq x - U_{n-1} + \mathfrak{D}_{n-1}} \mathrm{Bin}(\mathfrak{z}, \alpha, x - U_{n-1} + \mathfrak{D}_{n-1}) \\
&\quad \times \lambda_n^{\mathfrak{D}_n}(\mathfrak{z}, i, \ell) \mathrm{e}^{-\lambda_n(\mathfrak{z}, i, \ell)+1} \frac{\phi_n(U_n, \mathfrak{z}, i, \ell)}{\psi(U_n)} \rho_{n-1}(\ell, \mathfrak{z}) f_t.
\end{aligned}$$

Proof. We provide only the proof for $\mathfrak{q}_n(G_n^{m\ell i} D_n I(X_n = x))$.

First note that $G_n^{m\ell i} = G_{n-1}^{m\ell i} + \langle D_{n-1}, f_\ell\rangle\langle D_n, f_m\rangle\langle Y_{n-1}, e_i\rangle$. Therefore

$$
\begin{aligned}
&\mathfrak{q}_n(G_n^{m\ell i} D_n I(X_n = x)) \\
&= \overline{E}\big[\Gamma_n G_{n-1}^{m\ell i} D_n I(X_n = x) \mid \mathcal{Y}_n\big] \\
&\quad + \langle Y_{n-1}, e_i\rangle \overline{E}\big[\Gamma_n \langle D_{n-1}, f_\ell\rangle\langle D_n, f_m\rangle D_n I(X_n = x) \mid \mathcal{Y}_n\big] \\
&= \prod_{i,j=1}^{L} (La_{ji})^{\langle Y_n, e_j\rangle\langle Y_{n-1}, e_i\rangle} \sum_{l,t=1}^{N}\sum_{i=1}^{L} b_{tli}\langle Y_{n-1}, e_i\rangle \sum_{\mathfrak{z}\geq x - U_{n-1} + \mathfrak{D}_{n-1}} \\
&\quad \times \mathrm{Bin}(\mathfrak{z}, \alpha, x - U_{n-1} + \mathfrak{D}_{n-1})\lambda_n^{\mathfrak{D}_n}(\mathfrak{z}, i, l)\mathrm{e}^{-\lambda_n(\mathfrak{z}, i, l)+1}\frac{\phi_n(U_n, \mathfrak{z}, i, l)}{\psi(U_n)} \\
&\quad \times \overline{E}\Big[\big\langle \Gamma_{n-1} G_{n-1}^{m\ell i} D_{n-1} I(X_{n-1} = \mathfrak{z}), f_l\big\rangle \;\Big|\; \mathcal{Y}_{n-1}\Big] f_t + \prod_{i=1}^{L} (La_{ji})^{\langle Y_n, e_j\rangle\langle Y_{n-1}, e_i\rangle} \\
&\quad \times \sum_{i=1}^{L} b_{m\ell i}\langle Y_{n-1}, e_i\rangle \sum_{\mathfrak{z}\geq x - U_{n-1} + \mathfrak{D}_{n-1}} \mathrm{Bin}(\mathfrak{z}, \alpha, x - U_{n-1} + \mathfrak{D}_{n-1}) \\
&\quad \times \lambda_n^{\mathfrak{D}_n}(\mathfrak{z}, i, \ell)\mathrm{e}^{-\lambda_n(\mathfrak{z}, i, \ell)+1}\;\frac{\phi_n(U_n, \mathfrak{z}, i, \ell)}{\psi(U_n)} \\
&\quad \times \overline{E}\Big[\Gamma_{n-1}\langle D_{n-1}, f_\ell\rangle I(X_{n-1} = \mathfrak{z}) \;\Big|\; \mathcal{Y}_{n-1}\Big] f_\ell \\
&= \prod_{i,j=1}^{L} (La_{ji})^{\langle Y_n, e_j\rangle\langle Y_{n-1}, e_i\rangle} \sum_{l,t=1}^{N}\sum_{i=1}^{L} b_{tli}\langle Y_{n-1}, e_i\rangle \sum_{\mathfrak{z}\geq x - U_{n-1} + \mathfrak{D}_{n-1}} \\
&\quad \times \mathrm{Bin}(\mathfrak{z}, \alpha, x - U_{n-1} + \mathfrak{D}_{n-1}) \\
&\quad \times \lambda_n^{\mathfrak{D}_n}(\mathfrak{z}, i, l)\mathrm{e}^{-\lambda_n(\mathfrak{z}, i, l)+1}\frac{\phi_n(U_n, \mathfrak{z}, i, l)}{\psi(U_n)}\langle \mathfrak{q}_{n-1}(G_{n-1}^{m\ell i} D_{n-1} I(X_{n-1} = \mathfrak{z})), f_l\rangle f_t \\
&\quad + \prod_{j=1}^{L} (La_{ji})^{\langle Y_n, e_j\rangle\langle Y_{n-1}, e_i\rangle} \sum_{i=1}^{L} b_{m\ell i}\langle Y_{n-1}, e_i\rangle \sum_{\mathfrak{z}\geq x - U_{n-1} + \mathfrak{D}_{n-1}} \\
&\quad \times \mathrm{Bin}(\mathfrak{z}, \alpha, x - U_{n-1} + \mathfrak{D}_{n-1}) \\
&\quad \times \lambda_n^{\mathfrak{D}_n}(\mathfrak{z}, i, \ell)\mathrm{e}^{-\lambda_n(\mathfrak{z}, i, \ell)+1}\frac{\phi_n(U_n, \mathfrak{z}, i, \ell)}{\psi(U_n)}\rho_{n-1}(\ell, \mathfrak{z}) f_m.
\end{aligned}
$$

Remark 2. The filter recursions given above provide updates to estimate product processes, each involving processes D and X. What we would like to do, is manipulate these filters so as to remove the dependence upon these processes. This manipulation is routine.

$$
\begin{aligned}
&\sum_x \langle \mathfrak{q}_n(G_n^{m\ell i} D_n I(X_n = x)), \mathbf{1}\rangle = \sum_x \langle \overline{E}\big[\Gamma_n G_n^{m\ell i} D_n I(X_n = x) \mid \mathcal{Y}_n\big], \mathbf{1}\rangle \\
&= \sum_x \overline{E}\big[\Gamma_n G_n^{m\ell i} I(X_n = x)\langle D_n, \mathbf{1}\rangle \mid \mathcal{Y}_n\big] \\
&= \overline{E}\big[\Gamma_n G_n^{m\ell i} \sum_x I(X_n = x) \mid \mathcal{Y}_n\big] = \mathfrak{q}_n(G_n^{m\ell i}).
\end{aligned}
$$

It follows that

$$\mathfrak{q}_n(G_n^{m\ell i}) = \sum_x \langle \mathfrak{q}_n(G_n^{m\ell i} D_n I(X_n = x)), \mathbf{1}\rangle. \tag{8.12}$$

8.6 Parameter re-estimation

In this section, using the EM algorithm [7, 15, 24], the parameters of the model can be estimated.

Our model is determined by the set of parameters

$$\theta := (a_{ji},\ 1 \le i,j \le L,\ b_{m\ell i},\ 1 \le m,\ell \le N)$$

which we update to the new set

$$\hat{\theta} = \left(\hat{a}_{ji}(n),\ 1 \le i,j \le L,\ \hat{b}_{m\ell i}(n),\ 1 \le m,\ell \le N\right).$$

The following theorem is established using the techniques in either [5] or [16].

Theorem 3. *The new estimates of the parameters of the model given the observations up to time n are given, when defined, by*

$$\hat{a}_{sr}(n) = \frac{\mathfrak{T}_n^{rs}}{\mathfrak{J}_n^{r}}, \tag{8.13}$$

$$\hat{b}_{m\ell i}(n) = \frac{\mathfrak{q}_n(G_n^{m\ell i})}{\mathfrak{q}_n(S_n^{\ell i})}. \tag{8.14}$$

where the processes $\mathfrak{T}_n^{rs}$ and $\mathfrak{J}_n^{r}$ are defined by (8.8) and (8.9) respectively and $\mathfrak{q}_n(G_n^{m\ell i})$ and $\mathfrak{q}_n(S_n^{\ell i})$ are given by the recursions in Theorem 2 and Remark 2.

References

1. Aggoun, L., Benkherouf, L. and L. Tadj (1997). "A hidden Markov model for an inventory system with perishable items". *Journal of Applied Mathematics and Stochastic Analysis*, 10(4): 423–430.
2. Aggoun, L., Benkherouf, L. and L. Tadj (2000). "A stochastic jump inventory model with deteriorating items". *Stochastic Analysis and Application*, 18(1): 1–10.
3. Aggoun, L. and L. Benkherouf (2002). "M-ary detection of Markov modulated Poisson processes in inventory models". *Journal of Applied Mathematics and Computation,* 132: 315-324.
4. Aggoun, L., Benkherouf, L. and A. Benmerzouga (2002). "Recursive estimation of inventory quality classes using sampling". *Journal of Applied Mathematics and Decisions Sciences*, 7(4): 249–263.

5. Aggoun, L., and R.J. Elliott (2004). *Measure Theory and Filtering: Introduction and Applications*, Cambridge University Press, Cambridge, UK.
6. Al-Osh, M.N and A.A Alzaid (1987). "First order integer-valued autoregressive (INAR(1)) process". *Journal of Time Series Analysis*, 8: 261-275.
7. Baum, L.E. and T. Petrie (1966). "Statistical inference for probabilistic functions of finite state Markov chains". *Annals of the Institute of Statistical Mathematics*, 37: 1554-1563.
8. Alain, B., Çakanyildirim, M. and S. P. Sethi (2005). "Partially observed inventory systems: the case of zero balance walk". Working Paper SOM 200548, School of Management, University of Texas at Dallas,
9. Alain, B., Çakanyildirim, M. and S. P. Sethi (2005). "Optimal ordering policies for inventory problems with dynamic information delays". Working Paper, School of Management, University of Texas at Dallas, TX.
10. Alain, Çakanyildirim, M. and S. P. Sethi (2005). "On the optimal control of partially observed inventory systems". To appear in the Comptes Rendus de l'Academie des Sciences.
11. Alain, B., Çakanyildirim, M. and S. P. Sethi (2005). "Optimality of base stock and (s; S) policies for inventory problems with information delays". Working paper SOM 200547, School of Management, University of Texas at Dallas, TX.
12. Alain, B., Çakanyildirim, M. and S. P. Sethi (2005). "A multiperiod newsvendor problem with partially observed demand". Working Paper SOM 200550, School of Management, University of Texas at Dallas, TX.
13. Beyer, D., Sethi, S.P. and M.I. Taksar (1998). "Inventory models with Markovian demands and cost functions of polynomial growth". *Journal of Optimization Theory and Applications*, 98(2): 281-323.
14. DeCroix, G. A. and V.S. Mookerjee (1997). "Purchasing demand information in a stochastic-demand inventory", *European Journal of Operational Research* 102: 36-57
15. Dempster, A.P., Laird, N.M. and D.B. Rubin (1977). "Maximum likelihood from incomplete data via the EM algorithm". *Journal of the Royal Statistical Society*, B(39): 1-38.
16. Elliott, R. J., Aggoun, L. and J.B. Moore (1995). *Hidden Markov Models, Estimation and Control*, Springer Verlag.
17. Karaesmen F., Buzacott, J.A. and Y. Dallery (2002). "Integrating advance order information in make-to-stock production systems". *IIE Transactions* 34: 649-662.
18. McKenzie, E (1985). "Some simple models for discrete variate time series". *Water Res Bull*, 21: 645-650.
19. Monahan G.E. (1982). "A survey of partially observable Markov decision processes". *Management Science* 28(1): 1-16.
20. Nahmias, S. (1982). "Perishable inventory theory". *A review of Operations Research*, 30(4): 680-708.
21. Tan, T., Gullu, R. and N. Erkip (2003). "Optimal Inventory Policies under Imperfect Advance Demand Information".
22. Treharne, J.T. and C.R. Sox (2002). "Adaptive inventory control for nonstationary demand and partial information". *Management Science* 48: 607-624.
23. Yano, C.A. and H.L. Lee (1995). "Lot sizing with random yields: A review". *Operations Research* 43 (2): 311-333.
24. Wu, C.F.J. (1983) "On the convergence properties of the EM algorithm". *The Annals of Statistics*, 11: 95-103.

9

An empirical investigation of the unbiased forward exchange rate hypothesis in a regime switching market

Emilio Russo[1], Fabio Spagnolo[2] and Rogemar Mamon[3,4]

[1] Department of Mathematics, Statistics, Computing & Applications and Faculty of Economics and Business Administration
University of Bergamo, Italy
emilio.russo@unibg.it

[2] CARISMA and Department of Economics and Finance
Brunel Business School
Uxbridge, Middlesex, UK
Fabio.Spagnolo@brunel.ac.uk

[3] Department of Statistical and Actuarial Sciences
The University of Western Ontario
London, Ontario, Canada
rmamon@stats.uwo.ca

[4] CARISMA and Department of Mathematical Sciences
Brunel University
Uxbridge, Middlesex, UK

Summary. In this article we develop a model for exchange rate dynamics in an economy that exhibits regime shifts. The switching of regimes is modulated by a Markov chain in discrete time. A description of the foreign exchange market and of its stylised features is given. Finally, unbiased forward exchange rate hypothesis (UFER) is tested in the context of the US-dollar/UK-pound spot and forward exchange rates.

Key words: Exchange rate dynamics, structured change, unbiased forward exchange rate hypothesis

9.1 Introduction

Exchange rates are important variables in financial economics as they are essential inputs in the valuation of financial securities in the currency market. The economic and political mechanisms that generate exchange rates' changes over time and consequently the parameters or even the structure of the exchange rate model itself, may change as the economic and political environment changes. This kind of structural change was involved, for example, in the mechanism that generates the exchange rates under the European Monetary System (EMS). In fact, under the rules of the Exchange Rate Mechanism, central banks might intervene in currency markets to keep exchange rate within a target zone of pre-specified width. When it was believed that there would be a realignment in the near future, the exchange rates might become very volatile. When it seemed to be unlikely that a realignment would take place in the near future, however, there was expectation for different conditional distributions of the exchange rates. These different types of regimes were considered to give a motivation for the use of a regime-switching model and to characterise EMS exchange rates.

Modelling the conditional distribution of exchange rates as a regime-switching process is motivated by the occurrence of changes in monetary policy rules. For example, in the EMS target zone setting the economic motivation to use a regime-switching model was based on central bank policy regimes.

In the EMS situation, the dynamics of exchange rates were different from periods during which there was pressure on the weak currency and it was defended against speculative attack to periods during which the exchange rate bands were credible. The key difference between the US system and European system was the frequency of switches amongst regimes. In the US, regimes usually were long-lived and consequently regime switches were infrequent whilst in the EMS switches amongst regimes were quite frequent. In particular, episodes of extreme volatility speculative attack regime, where the weak currency was defended by the central bank, did not tend to last long. In this regime, a speculative attack or a change in the fundamentals drove the exchange rate towards the weak edge of the target zone. The central bank of the depreciating currency might intervene in foreign exchange markets or raise interest rates to drive the exchange rate back towards the center of the target zone. Sometimes, central banks were successful in averting the currency crisis but, sometimes they were not, so a realignment might occur. Clearly, as the US regime tended to be more long-lived than the EMS regime that was more volatile, the number of switches amongst regimes was likely to be larger for EMS exchange rates than US rates.

In the literature, many models have Markov regime switching to describe the behaviour of economic variables subject to structural changes, both in stationary environments (see Kim and Nelson [32] for further details) and nonstationary ones (see Hall et al. (1997), Paap and van Dijk [38], Psaradakis et al. [39] amongst others). Statistical inference for these models is likelihood-based exploiting the fact that the maximum likelihood estimator is consistent.

Many researchers devoted their efforts studying the expectation hypothesis for interest and exchange rates in the context of Markov switching models. For example, in the related area of interest rate theory, Hamilton [28], Lewis [35], Driffill [11], Sola and Driffill [40], Evans and Lewis[17], [18], Evans [16], Gray [25], and Ang and Bekaert [1] investigated whether structural breaks can account for the rejection of the expectation hypothesis, assuming that the stochastic process which generates the short term interest rates is subject to Markov regime shifts.

Even though there is a consensus that US interest rates are best described as processes that are subject to changes in regime, the results concerning the empirical validity of the expectation hypothesis are far from conclusive, with many studies reporting evidence which is not consistent with the predictions of the expectation theory. Several authors noted that the explanatory power of the expectation hypothesis of the term structure tends to be greater outside the United States. Hardouvelis [30] examined the behaviour of three-month and ten-year rates in the G-7 countries and found that the expectation hypothesis works particularly well for all the countries but the United States. Further evidence suggests that it is more difficult to reject the expectation hypothesis using non-US data as Gerlach and Smets [23] provided.

Considerable instability of term structure regressions is also documented by Dahlquist and Jonsson (1995), who supplied evidence that the expectation hypothesis can be rejected for periods in which foreign exchange market were calm.

In this paper we present an econometric model which allows for the presence of a Markov regime-switching behaviour of exchange rate series. In section 2 we analyse the stylised features and the statistical properties of foreign exchange rate. We concentrate our attention on spot and forward US-dollar/UK-pound exchange rate series. In section 3 we provide a theoretical survey concerning the stationarity property of the series whilst in section 4 we discuss cointegration and use an error correction model to test if the spot and forward exchange rate series are well modelled by the UFER. In section 5 we consider a simple Hamilton model to capture the changes in regime governed by a Markov chain dynamics and identify the turning point of the US-dollar/UK-pound spot and forward time series. Finally, in section 6 we give some concluding remarks.

9.2 Stylised features and statistical properties of foreign exchange rates

The purpose of this section is to analyse the so-called stylised features of the exchange rate time series, that is to examine the empirical regularities which are commonly found in the exchange rates dynamics. In fact, there is a number of well known (and less well known) stylised features about the empirical behaviour of exchange rates and not all of them are considered in empirical and theoretical economic research (see de Vries [7] for further details). Most of the series may contain a stochastic trend and the natural consequence of this fact is that these series do not have time-invariant movements and are therefore nonstationary. The nonstationary property of the series is also evident from the behaviour of the series themselves that seem to meander. This implies that any shock to a series displays a high degree of persistence and the volatility of many series is not constant over time[5]. Sometimes, some time series share movements that are common to other time series so it is possible to relate the series with each other. This is what happens when we consider the spot and the forward exchange rate time series.

We focus our analysis principally on monthly US-dollar/UK-pound exchange rate series by studying its behaviour and by providing a statistical treatment of its stylised features. Then, a Markov regime-switching model is used to test if such model is able to capture very well the behaviour of the considered exchange rate series.

It is worth noting that many empirical studies, concerning the behaviour of exchange rates, have appeared in the last thirty years. Amongst the many relevant works are the study of Mussa [37] showing the empirical regularities in the behaviour of exchange rates, followed by the work of Levich [34] that explains the price behaviour of related rates, and that of Frenkel and Meese [21] on the variability of the exchange rates. A comprehensive analysis of the econometrics of exchange rates are also provided in Taylor [42], Diebold [10], and Baillie and McMahon [2].

We investigate the relationship between spot and forward rates and the implication of this relationship on testing the UFER hypothesis. There is an enormous literature on testing whether the forward exchange rate is an unbiased predictor of future spot exchange rate. As in the approach proposed by Cornell [5] Levich [33] and Frenkel (1980), we base our analysis on the regression of the logarithm of the future spot rate, s_{t+1}, and the logarithm of the current forward rate, f_t. Then, related to the studies of Bilson [4], Fama [19], and Froot and Frenkel [22], we concentrate on the regression of the change in the logarithm spot rate, Δs_{t+1}, on the forward premium, $f_t - s_t$. To conduct these analyses, we consider two variables of interest: spot and forward foreign exchange rates. The spot rate, which is the exchange rate quoted for immediate delivery of the currency of the buyer, is seen as the best variable that can help us analyse trade-related problems. The spot exchange rate is the variable that clears the market for exports and imports. The forward rate, which is the guaranteed price agreed today at which the buyer will take delivery of

[5] Generally, periods of low volatility are followed by periods in which the volatility is relatively high.

currency at some future periods, is very well related to the spot rate as we shall show.

Our analysis is based on spot, displayed in Figure 9.1 and forward displayed in Figure 9.2, monthly foreign exchange rates US-dollar/UK-pound during the period October 1983-January 2005. The exchange rates seem to go through periods of appreciation and then depreciation as we can see from their plots. Both the US-dollar/UK-pound spot rates graph and the forward rates one, may be divided in two well separated subperiods. The first period, characterised by a higher variance until 1993, is then followed by a second period characterised by a lower variance whilst the mean level is similar in the two subperiods.

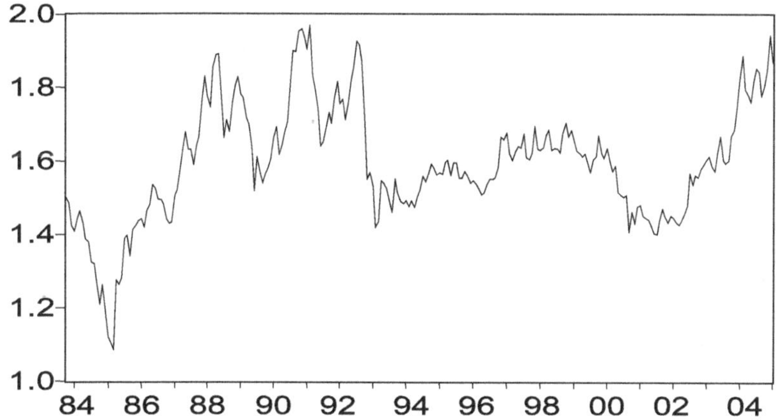

Fig. 9.1. Plot of US-dollar/UK-pound spot exchange rates during the period October 1983-January 2005

The summary statistics of the entire set of data are given in Table 9.1. We use spot and forward nominal exchange rate to calculate the statistics of the series. From Table 9.1, it is apparent that the spot and forward series have characteristics that are very similar to each other. Hence, we can explain, for example, the spot series by using the forward series and vice-versa. In fact, there are many theoretical and empirical works that focus on the analysis of the UFER hypothesis, which asserts that the forward exchange rate is an unbiased predictor of the future spot exchange rate.

Returning to the observations related to the series' behaviour, the first step is to divide the spot and forward historical series in two or more subperiods to check if in the subperiods there is a change in mean and variance that imply a change in the behaviour of the exchange rates. We choose only two different subperiods since there is no evidence of other well distinguished

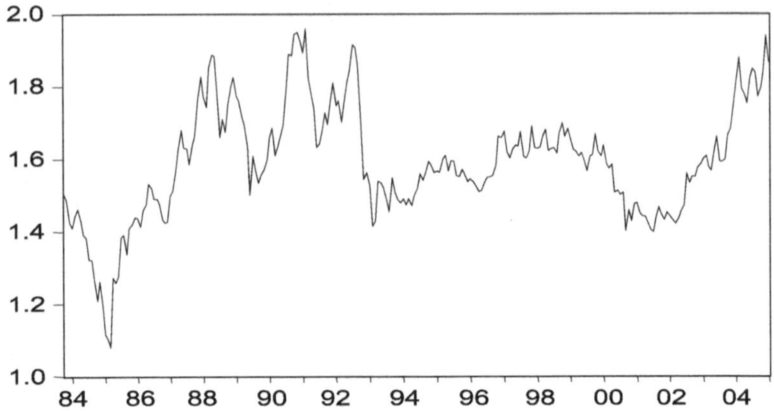

Fig. 9.2. Plot of US-dollar/UK-pound forward exchange rates during the period October 1983-January 2005

Statistics	*Spot*	*Forward*
Mean	1.593557	1.590145
Median	1.591250	1.586950
Maximum	1.970000	1.960400
Minimum	1.084000	1.079100
Standard Deviation	0.159323	0.158029
Skewness	-0.078759	-0.104196
Kurtosis	3.453549	3.479265
Number of observations	256	256

Table 9.1. Summary statistics for spot and forward US-dollar/UK-pound exchange rates 1983-2005

subperiods. For US-dollar/UK-pound series the first subperiod is considered from October-1983 to December-1993 whilst the second is considered from January-1994 to January-2005. Tables 9.2 and 9.3 give the descriptive statistics for spot and forward series in the two specified subperiods. In the US-dollar/UK-pound subperiods we note a similar mean level, that changes from 1.595042 to 1.592184 in the spot series whilst from 1.589986 to 1.590292 in the forward one; however, the standard deviation of the first subperiod is almost two times the standard deviation of the second subperiod for both spot and forward series. Thus, the two subperiods are characterised by a strong change in variance. These observations give support to the presence of two well separated regimes and of the fact that there is an evident change in the behaviour of the exchange rates from one subperiod to another. One of them is characterised by a higher level of variance whilst the other is characterised by a lower level of variance. This justifies the use of a regime-switching Markov

Statistics	*Spot*	*Forward*
Mean	1.595042	1.589986
Median	1.590000	1.585400
Maximum	1.970000	1.960400
Minimum	1.084000	1.079100
Standard Deviation	0.200074	0.198361
Skewness	-0.225703	-0.233914
Kurtosis	2.566956	2.580891
Number of observations	123	123

Table 9.2. Summary statistics for the US-dollar/UK-pound exchange rate series (October 1983-December 1993)

Statistics	*Spot*	*Forward*
Mean	1.592184	1.590292
Median	1.592500	1.587500
Maximum	1.944050	1.940200
Minimum	1.400700	1.398400
Standard Deviation	0.109709	0.108984
Skewness	0.710399	0.697703
Kurtosis	3.656456	3.648736
Number of observations	133	133

Table 9.3. Summary statistics for the US-dollar/UK-pound exchange rate series (January 1994-January 2005)

model to capture this type of behaviour in the exchange rates series. But first, we focus our attention on the characteristics of the two series by testing the stationarity or nonstationarity of the spot and forward series.

9.3 Stationary and nonstationary time series

Time series may be stationary, trend stationary or nonstationary. A stationary time series has a constant mean, a constant variance and the autocovariance results to be independent of time. Stationarity is one of the more desirable properties in standard econometric theory as, without stationarity, it is not possible to obtain consistent estimators for the considered time series. Plotting the series against time is a quick way to check if a series is stationary. If the graph crosses the mean many times, this signals that the series in question is stationary. Conversely, if the graph shows the opposite situation, this indicates persistent trends away from the mean of the series. A trend stationary series is a series whose mean grows around a fixed trend. It means that a trend-stationary series tends to evolve around an upward sloping curve without big swings away from that curve. The concept of nonstationary time series is strictly related to the concept of unit root. A unit root process is characterised by an infinite variance and will only cross the mean of the sample somewhat infrequently. Furthermore, it displays very long positive or negative strays away from the sample mean. A process that has a unit root is also called integrated of order one, denoted by $I(1)$. On the other hand, a stationary process is integrated of order zero, denoted by $I(0)$.

There are important differences between stationary and nonstationary time series. Stationary series are characterised by shocks that vanish over time and the series revert to their long run mean level. Consequently, a long term forecast converges to the unconditional mean of the series. Furthermore, stationary series exhibit mean reversion in that they fluctuate around a constant long-run mean, give a finite variance that is time-invariant and have a theoretical autocorrelogram that diminishes as lag length increases. On the other hand, nonstationary series have mean and variance that are time-dependent; thus, they are identified by the lack of a long-run mean to which the series return. A sample correlogram could be used to detect for the presence of unit root but it is qualitative and would be an imprecise approach to use in testing the null hypothesis of a unit root. Consider the first-order process

$$y_t = ay_{t-1} + \epsilon_t.$$

Dickey and Fuller [8], [9] developed a procedure to test for the presence of a unit root for the process above. Their methodology is based on the generation of thousands of random walk sequences and for each one of them they estimate the value of a. Although most of these calculated values are near to unity, some would be further from unity than others. By using this procedure, Dickey and Fuller found critical values to test for unit roots. Stationarity necessitates the condition $|a| < 1$. Thus, if the estimated value of a is close to 1, we should be concerned about nonstationarity. If we define $a' = a - 1$, the equivalent stationarity condition is $-2 < a' < 0$. The Dickey-Fuller test may be conducted to check that the estimated value of a' is greater than -2. The procedure that Dickey and Fuller used to determine their critical values, is typical of that found in modern time series analysis and is related to the fact that hypothesis tests, concerning the coefficients of nonstationary variables, cannot be conducted by using the standard t-test or F-test. The distributions of the appropriate test statistics are not in analytic forms and cannot be evaluated analytically. The evaluation of these non-standard distributions may easily be carried out using a Monte-Carlo simulation. We use the Dickey-Fuller (DF) test to look for evidence of the presence of a unit root.

Let s_t be the logarithm of spot exchange rate value at time t. We define the first difference as

$$\Delta s_t = s_t - s_{t-1}.$$

Guided by the methodology of Dickey and Fuller [8], we consider three different regression equations that can be used to test for the presence of a unit root. For spot series these are

$$\begin{aligned} \Delta s_t &= \alpha s_{t-1} + \epsilon_t, \\ \Delta s_t &= c + \alpha s_{t-1} + \epsilon_t, \\ \Delta s_t &= c + \alpha s_{t-1} + bt + \epsilon_t. \end{aligned} \tag{9.1}$$

The regression equations in (9.1) may easily be defined also for the logarithm of forward exchange rate, f_t, at time t. The difference between the three regressions concerns the presence of the deterministic terms c and bt. The first is a pure random walk model, the second adds an intercept whilst the third adds both an intercept and a linear time trend. The parameter of interest in all the regression equations is α. If $|\alpha| = 0$ there is evidence of the presence of a unit root. The test involves each equation above by using the ordinary least squares (OLS) method in order to obtain the estimated value of α and associated standard error. The comparison between the resulting t-statistics and the appropriate value reported in the Dickey-Fuller tables allows us to determine whether to accept or reject the null hypothesis $|\alpha| = 0$. We can conduct the same type of analysis by replacing (9.1) by the autoregressive process as we can find in the augmented Dickey-Fuller test (ADF). According to this methodology, we can model the first difference of the logarithm of spot and forward series through the equation

$$\Delta s_t = c + \alpha s_{t-1} + \beta \Delta s_{t-1} + \gamma \Delta s_{t-2} + \delta \Delta s_{t-3} + \varepsilon \Delta s_{t-4} + ... + \epsilon_t, \tag{9.2}$$

where c is a constant whilst $\alpha, \beta, \gamma, \delta$ and ε are the coefficients associated with the past data and ϵ_t is white noise. In equation (9.2) the coefficient of interest is α; if $|\alpha| = 0$ the equations are entirely in first differences and, consequently, it has a unit root.

The Dickey-Fuller test assumes that the errors are independent and have a constant variance. This raises the important issue that we do not know the true data-generating process. We cannot estimate properly α and its standard error unless all the autoregressive terms are included in the estimating equation. Since the true order of the autoregressive process is unknown, for both spot and forward exchange rate, the problem now is to select the appropriate lag length. Including too many lags reduces the power of the test to reject the null of a unit root since the increased number of lags necessitates the estimation of additional parameters and a loss of degree of freedom. In contrast, too few lags will not appropriately capture the error process so that α and its standard error cannot be estimated correctly. We start with a relatively long lag length and then we decrease the lag length by considering the t-test value. The process is repeated until the lag is significantly different from zero. We definitively choose 4 lags to proceed with our analysis, as they are necessary to have no correlation in the residuals.

Table 9.4 illustrates the value of the ADF-test for our exchange rate series data. We consider the regression equation in (9.2) without considering a linear time trend. Comparison of the resulting ADF test statistics with the critical values supplied by Dickey and Fuller shows that there is evidence that the logarithm series of spot and forward exchange rates US-dollar/UK-pound are significantly nonstationary at 90%, 95% and 99% level.

ADF Test Statistics		*Critical Value*	*Level Critical Value*
Spot	-2.403479	-3.4580	1%
		-2.8731	5%
Forward	-2.411600	-2.5729	10%

Table 9.4. Unit root test results for US-dollar/UK-pound spot and forward exchange rates

9.4 Cointegration and the unbiased forward exchange rate (UFER) hypothesis

Given the nonstationarity of the considered spot and forward exchange rate series, it is possible that a linear combination of integrated variables is stationary. Such variables are said to be cointegrated. Zivot [43] gave additional discussion and details of the concept of cointegration within the context of forward and spot exchange rate regressions.

Cointegration was a terminology introduced by Engle and Granger [15]. We present their formal analysis following Enders [12]. Suppose to consider a set of economic variables in long-run equilibrium when

$$\beta_1 x_{1t} + \beta_2 x_{2t} + ... + \beta_n x_{nt} = 0.$$

If we let β and $\mathbf{x_t}$ denote the columns vectors of the equilibrium constants $(\beta_1, \beta_2, ..., \beta_n)$ and of the variables $(x_{1t}, x_{2t}, ..., x_{nt})$ respectively, the system is in long-run equilibrium if $\beta^{'}\mathbf{x_t} = 0$, where $\beta^{'}$ denotes the transpose of the vector β. The deviation from long-run equilibrium, called equilibrium error, is $\mathbf{e_t}$ so that $\mathbf{e_t} = \beta^{'}\mathbf{x_t}$. Engle and Granger (1987) provided the following definition of cointegration: the components of the vector $\mathbf{x_t} = (x_{1t}, x_{2t}, ..., x_{nt})$ are said to be cointegrated of order (d, b), denoted $x \sim CI(d, b)$, if

- all components of $\mathbf{x_t}$ are integrated of order d;
- there exists a vector $\beta = (\beta_1, \beta_2, ..., \beta_n)$ such that the linear combination $\beta\mathbf{x_t} = \beta_1 x_{1t} + \beta_2 x_{2t} + ... + \beta_n x_{nt}$ is integrated of order $(d - b)$, where $b > 0$. The vector β is called a cointegrating vector.

According to the definition provided by Engle and Granger [15], cointegration refers to a linear combination of nonstationary variables but it is quite possible that a non-linear combination may exist amongst integrated variables.

Then, all variables must be integrated of the same order. Of course, this does not imply that all similar integrated variables are cointegrated. As a matter of fact, a set of $I(d)$ variables is usually not cointegrated. Such a lack of cointegration implies no long-run equilibrium amongst the variables, so that they can wander arbitrarily far from each other. If the variables are integrated of different orders, they cannot be cointegrated.

Engle and Granger [15] proposed a test that determines whether or not two $I(1)$ variables are cointegrated of order $CI(1,1)$. To explain this testing procedure for cointegration, consider two variables y_t and z_t integrated of order 1, $I(1)$, and the following steps are carried out.

Step 1. Pretest the two variables for their order of integration. A DF-test (or an ADF-test) may be used to test for the order of integration. By definition, cointegration necessitates that the variables be integrated of the same order. If the variables are integrated of different order, we can conclude that they are not cointegrated.

Step 2. Estimate the long-run equilibrium relationship. If the results provided in Step 1 indicate that the two variables are $I(1)$, the next step is to estimate the long-run equilibrium relationship in the form

$$y_t = \alpha + \beta z_t + e_t. \tag{9.3}$$

In order to determine if the variables are cointegrated, we have to consider the residual sequence of the long-run equilibrium relationship. Let $\hat{e}_t$ denote the series of the estimated residuals of the long-run relationship

$$\hat{e}_t = y_t - \hat{\alpha} - \hat{\beta} z_t,$$

where $\hat{\alpha}$ and $\hat{\beta}$ are the estimated values of the parameters α and β in (9.3). If this last series is stationary, y_t and z_t are cointegrated of order (1,1), symbolically $CI(1,1)$. It is convenient to perform a DF-test on these residuals to determine their order of integration and consequently to find out if they are stationary. Consider the autoregression of the residuals

$$\Delta \hat{e}_t = a_1 \hat{e}_{t-1} + \epsilon_t. \tag{9.4}$$

The parameter of interest is a_1. If we cannot reject the null hypothesis $|a_1| = 0$, we can conclude that the residuals series contains a unit root and, consequently, y_t and z_t are not cointegrated. Conversely, the rejection of the null hypothesis $|a_1| = 0$ implies that the residual sequence is stationary, and given that both y_t and z_t are $I(1)$, we can conclude that y_t and z_t are cointegrated of order (1,1). If the residuals in (9.4) do not appear to be white noise, an ADF-test may be used. Suppose the series ϵ_t of (9.4) exhibits serial correlation. Instead of considering (9.4), estimate the autoregression

$$\Delta \hat{e}_t = a_1 \hat{e}_{t-1} + \sum_{i=1}^{n} a_{i+1} \Delta \hat{e}_{t-i} + \epsilon_t.$$

Again, if we can reject the hypothesis $| a_1 |= 0$, we conclude that the residual sequence is stationary and y_t and z_t are $CI(1,1)$.

The relationship between cointegration and the UFER hypothesis has been discussed by many authors starting with Hakkio and Rush [26]. Engel [13] provided a survey of this literature. The UFER is one form of the efficient market hypotheses and asserts that the forward price of an asset should equal the expected value of that asset's spot price in the future. Related to this is the fact that forward exchange market efficiency requires the one-period forward exchange rate equal the expectation of the spot rate in the next period. If f_t represents the logarithm of the one-period price of forward foreign exchange rate at time t and s_t is the logarithm of the spot foreign exchange rate at the same time, the UFER hypothesis asserts that, under rational expectation and risk neutrality, it must be the case that

$$E_t[s_{t+1}] = f_t,$$

where $E_t[\cdot]$ is the expectation conditional on information available at time t. If this relationship fails, speculators can expect to make a pure profit of their trades in the foreign exchange market. If the agents' expectations are rational, the forecast error for the spot rate at time $t+1$ is characterised by zero conditional mean, so that

$$s_{t+1} - E_t[s_{t+1}] = \varepsilon_{t+1},$$

where ε_{t+1} is a random variable called rational expectation forecast error with $E_t[\varepsilon_{t+1}] = 0$. Combining the last two equations yields

$$s_{t+1} = f_t + \varepsilon_{t+1}. \tag{9.5}$$

Two different regression equations have been used to test the UFER. The first is the level regression

$$s_{t+1} = a_0 + a_1 f_t + \epsilon_{t+1}. \tag{9.6}$$

The null hypothesis that UFER is true imposes the restrictions $a_0 = 0$, $a_1 = 1$ and $E_t[\epsilon_{t+1}] = 0$. Since the unit root test gives evidence that spot and forward exchange rates are both integrated of order one, $I(1)$, the UFER hypothesis requires that s_{t+1} and f_t must be cointegrated with cointegrating vector (1,-1) and that the stationary cointegrating residuals, ϵ_{t+1}, satisfy $E_t[\epsilon_{t+1}] = 0$. It means that there should be a linear combination of nonstationary spot and forward exchange rates that is stationary.

Meese and Singleton [36] and Isard [31] provided an explanation of the fact that, since s_t and f_t have unit roots, the level equation (9.6) is not a valid regression equation because of the spurious regression problem described by Granger and Newbold (1974). In fact, the spurious regression presents a t-statistic that appears to be significant but the results are without any economic meaning (see Granger and Newbold [24]).

The second regression equation used to test the UFER hypothesis is the difference equation

$$\Delta s_{t+1} = \alpha + \beta(f_t - s_t) + \epsilon_{t+1}. \tag{9.7}$$

In fact, the aim behind cointegration is the detection and analysis of long-run relationships amongst time series variables. Given that spot and forward time series appear to be nonstationary they often require differencing to be trasformed into stationary. The problem with differencing, however, is that this may remove relevant long-run information. The cointegration analysis provides a way of retaining both short-run and long-run information. As spot and forward exchange rate series are both $I(1)$, to have all variables in the regression integrated of the same order for equation (9.7), the forward premium $f_t - s_t$ should be $I(0)$ or, equivalently, f_t and s_t should be cointegrated of order (1,-1). At first, we use an error correction model (ECM) to test if the spot and forward exchange rate series are well modelled by using the UFER hypothesis. By subtracting s_t from both sides of (9.5), we have

$$\Delta s_{t+1} = f_t - s_t + \epsilon_{t+1}.$$

The ECM is thus defined exactly as the difference equation (9.7) and the null hypothesis that UFER is true, imposes the restrictions $\alpha = 0$, $\beta = 1$ and $E_t[\epsilon_{t+1}] = 0$.

The first step is to test if the two considered series Δs_t and $f_t - s_t$ are stationary. We use again an ADF test. The results are presented in Table 9.5. The

ADF Test Statistics		*Critical Value*	*Level Critical Value*
Δs_t	-14.95459	-3.4577	1%
		-2.8730	5%
$f_t - s_t$	-9.011005	-2.5728	10%

Table 9.5. ADF-test results for US-dollar/UK-pound series

series are stationary at confidence levels of 90%, 95% and 99% and present a high ADF test statistic in absolute value with respect to the critical values.

We are now in a position to consider the ECM to test the UFER hypothesis. This means that we want to test the hypothesis $\alpha = 0$ and $\beta = 1$. We use the Wald test for this purpose. According to our analysis and to the results presented in Tables 9.6 and 9.7, the Wald test results to a rejection of the UFER hypothesis for US-dollar/UK-pound series. We note that when the UFER hypothesis is rejected, the typical estimate of β, $\hat{\beta}$, is significatively negative. This result is often referred to as the forward discount anomaly or forward discount bias but the principal reason why $\hat{\beta}$ is negative, is due to the unaccounted regime shifts in modelling the behaviour of the considered exchange rate series. All these observations led us to consider a Markov regime-switching model to capture the shifts of the US-dollar/UK-pound exchange rate series (see Driffill [11], Zivot [43], Spagnolo, Psaradakis and Sola [41]).

Spagnolo, Psaradakis and Sola [41] offered a possible explanation for the empirical evidence of the rejection of the UFER hypothesis. In particular, they exploited an implication of the consumption capital asset pricing model under structural changes in consumption to reconcile this empirical evidence with general equilibrium models. The motivation for this approach relies on the empirical finding that consumption dynamics can be characterised by models that allow for structural changes that are driven by a Markov process. When combined with the hypothesis of time-varying risk-premium, such dynamic behaviour for consumption implies that the risk-premium itself is subject to Markov changes in regime. The presence of a Markov switching risk-premium further leads the authors to use a model for the spot rate and the forward premium whose parameters switch stochastically amongst regimes.

Dependent variable	Δs_{t+1}	
Coefficients	*Value*	*Std. Err.*
$\hat{\alpha}$	-0.002731	0.002598
$\hat{\beta}$	-1.714253	0.842843

Table 9.6. ECM estimations for US-dollar/UK-pound series

Null hypothesis: $\alpha = 0; \beta = 1$		
Statistics Test	Value	Probability
F-statistic	6.389142	0.001963
Chi-square	12.77828	0.001680

Table 9.7. Wald test results for US-dollar/UK-pound series

9.5 Evidence from exchange rate market via a Markov regime-switching model

The results obtained in the previous sections may be explained by considering a Markov regime-switching model to capture the presence of the evident change in variance that affects the considered historical data. We consider a simple Hamilton model to capture the changes in regime by using the Markov's hypothesis to identify the turning point of the US-dollar/UK-pound spot and forward time series.

As in Hamilton's work [29], we consider a two-state economy where the unconditional mean and variance are the only two parameters that can identify a change in regime. The idea is to apply this model to explain the behaviour of the exchange rate series and we argue that an explanation for the results presented above may lie with the forward premium $f_t - s_t$ subject to discrete Markov shifts. To investigate this possibility, we first investigate the property of the series $f_t - s_t$, as we have done in the previous sections. In fact, any Markov-type nonlinearities in the forward premium are likely to be reflected in the dynamic behaviour of the first difference spot series, Δs_{t+1}.

In this section we consider a system with two regimes where the regime identification variable, denoted by X_t, may only assume two possible values, 0 or 1. We suppose that the transition between two regimes follows a first-order Markov process. Moreover, let p denote the probability $P[X_t = 0 \mid X_{t-1} = 0]$ whilst q denote the probability $P[X_t = 1 \mid X_{t-1} = 1]$. We choose the underlying model as the ECM defined by the equation

$$\Delta s_{t+1} = \alpha + [\beta_0 + (\beta_1 - \beta_0)X_t](f_t - s_t) + [\sigma_0 + (\sigma_1 - \sigma_0)X_t]\epsilon_{t+1}, \quad (9.8)$$

where σ_i, for $i = 0, 1$, identifies the standard deviation of the two regimes. Furthermore, we consider a regime-switching Markov model that allows both the parameter β and σ to switch between two values according to a time-homogeneous Markov transition process. By using the Hamilton's procedure (1989), we forecast the forward premium series $f_t - s_t$ of the exchange rate US-dollar/UK-pound series. This is obtained through an optimal inference concerning the current state given the past value of the series $(f_{t-1} - s_{t-1}), (f_{t-2} - s_{t-2}), \ldots$. This first step is provided by the Hamilton [28],[29] non-linear filter according to which it is possible to infer the historical sequence of states X_t by considering the observed sequence of data $f_t - s_t$. All the parameters involved in the analysis are estimated by maximum likelihood. The sample likelihood function is given as a byproduct of the filter. It is then maximised numerically with respect to all the parameters and subject to the constraint that p and q lie in the open unit interval. The second step is to use the outcome of the filter to generate future forecasts of the series $f_t - s_t$.

As illustrated in the previous sections, the variance of the US-dollar/UK-pound series seems to switch between regimes evident from a significant difference between different subperiods. Thus, in light of all these observations, we proceed to model the first difference in the spot series s_t under the assumption that the forward premium $f_t - s_t$ is subject to Markov changes in regime. In (9.8) the UFER hypothesis is equivalent to the specifications $\alpha = 0$

and $\beta_i = 1$, $i = 0, 1$. We employ the recursive algorithm described in Hamilton [29] to estimate and test Markov switching model in (9.8), where the error term ϵ_{t+1} and the Markov chain X_t are not observed.

Tables 9.8, 9.9 and 9.10 provide the results of the analysis conducted on the market data on US-dollar/UK-pound exchange rate series. These results are obtained by using the software Gauss together with the Hamilton's [29] filter. We note that the estimated transition probabilities p and q are near unity.

Parameters	*Value*	*Standard error*
p	0.8696	0.0834
q	0.9624	0.0226
α	-0.0018	0.0022
β_0	-2.4094	0.7940
σ_0	0.0227	0.0017
σ_1	0.0472	0.0066

Table 9.8. Regime-switching model parameter estimates for the US-dollar/UK-pound series

	Q-statistics	*Probability*
Q(1)	0.6088	0.4352
Q(6)	1.3796	0.9671
Q(12)	7.6624	0.8109

Table 9.9. Box-Pierce Q-statistics on standardised residuals for the US-dollar/UK-pound series

	Q-statistics	*Probability*
$Q^2(1)$	0.0628	0.8021
$Q^2(6)$	7.8230	0.2514
$Q^2(12)$	10.4208	0.5791

Table 9.10. Box-Pierce Q-statistics on squared standardised residuals for the US-dollar/UK-pound series

This suggests that the Markov chain driving the changes in regime is highly persistent, so if the system is in either of the two regimes, it is likely to remain in that regime for a long time. Furthermore, the two regimes are distinct, with the standard deviation of exchange rate changes being two times higher in one regime than the other[6]. Table 9.8 confirms that for the considered series there is evidence of two well-specified regimes. The first regime is characterised by a negative high level of the parameter β_0 that has a value of -2.4094 whilst the second regime is characterised by a positive value of 0.5585 for the β_1 parameter. In accordance to the theory, we note further that the levels of standard deviation in the two regimes are significantly different. In particular, the standard deviation associated with regime 0 is 0.0227 whilst the standard deviation associated with regime 1 is 0.0472. In our analysis we use a constant intercept model. By considering different values for the intercept, our analysis does not give evidence of significant different results. Finally, the resulting Q-statistics, based on standardised residuals and squared standardised residuals, provide evidence of the absence of autocorrelation amongst residuals. All these observations allow us to identify two different regimes and motivate the use of a regime-switching Markov model to capture the behaviour of the series that switches from one regime to another.

Plot of the estimated filtered probabilities, in Figure 9.3, shows evidence of two distinct regimes. The first regime corresponds to the first subperiod, from October 1983 to December 1993, covering until the $123rd$ observation. The second regime is identified by the second part of the graph, from January 1994 to January 2005, where we observe a new behaviour in the exchange rate US-dollar/UK-pound series. During the periods under examination, there are several changes in regime. For example, the filter allocates a high probability for state $X_t = 1$ in the first subperiod of the early 1990s, specifically up to roughly the $100th$ observation. This is probably caused by the UK's exit from the ERM following disruption on the financial markets.

The economic motivation for the presence of two distinct regimes, the first until 1993 and the second beginning after 1993, is related to the international monetary relations that have exhibited cycles during the last half century. Examples of these cycles are represented by the breakdown of the Bretton Woods regime in early 1970s or the conflicts over world reflation[7] that were resolved at the Bonn summit in 1978. But, what has a direct influence on the exchange behaviour of the US-dollar against the UK-pound is the recession and the consequent recovery that happens during 1990s. In fact, the US economy experienced a recession in 1991 and a slow recovery in 1992 with only a delayed response in the market. It was only in January 1993 that the US economy was beginning to recover whilst those of Europe and Japan lagged behind; this poised for export-led growth at the expense of the US current account position. Another important turning point is represented by the decision of EU, in the 1990s, to form a monetary union that could represent a

[6] This is consistent with the long-swings in the dollar reported by Engel and Hamilton [14] and Bekaert and Hodrick [3].

[7] This refers to the intentional reversal of deflation through a monetary action of a government.

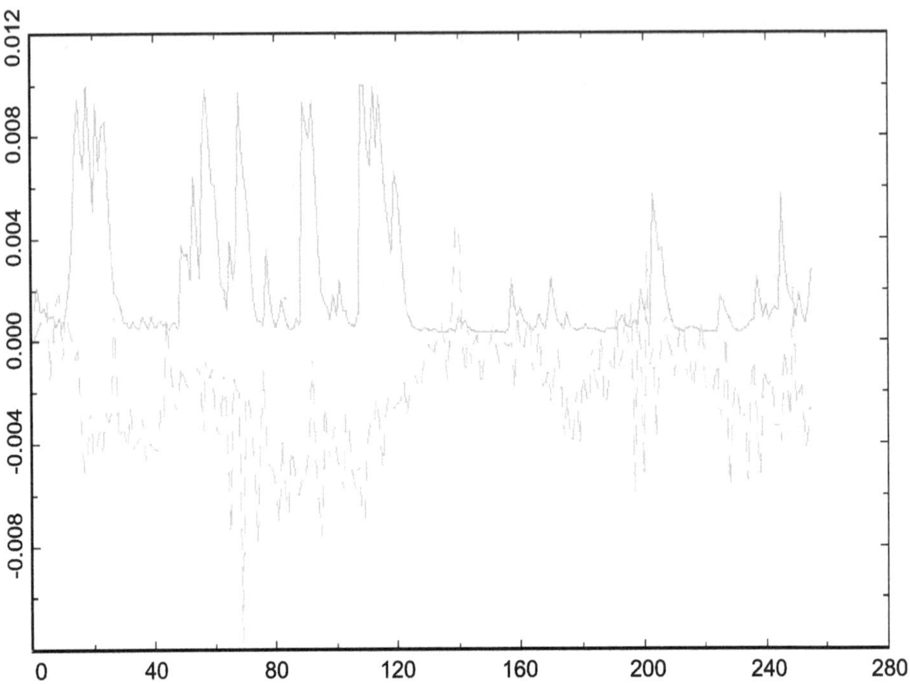

Fig. 9.3. Plot of filtered probabilities (upper half) and $(f_t - s_t)$ graph (lower half) for the logarithm of US-dollar/UK-pound exchange rate series for all the 256 available monthly data

counterweight to United States and its dollar in the international monetary system. This initiation leads to a situation in which European states have more power and are less susceptible to pressure from the United States for policy change and to fluctuations in the US dollar. Thus, over the long-run, the structure of the system of the exchange rates could respond to the policy behaviour of the dominant state. As noted previously an evidence of a regime change in the behaviour of US-dollar/UK-pound series leads to a situation that happened in 1993.

To summarise, our analysis has been based on a theoretical econometric model which allows for the presence of a Markov regime-switching behaviour of exchange rate series. Under these conditions, the forward premium itself is subject to changes in regime. As a consequence, a model for the spot rate and forward premium is characterised by parameters that depend on the state of economy and explanatory variables that are correlated with the disturbances. We have provided evidence that a regime-switching Markov model is able to capture the changes in regime present in the exchange rate behaviour of the data studied in this paper.

9.6 Concluding remarks

The use of a regime-switching model to characterise the behaviour of exchange rate is justified by the on-going changes that may happen in the economic and political environment. This modelling framework is intimately connected with the Hamilton's Markov model that capture the changes in regime of a financial time series. The idea is to consider a discrete-time Markov process to model the switches in regime. Although the model here is initially based on two regimes, this can be extended to more than two regimes.

Preliminary analysis of the statistical features of the data series clearly signifies the success of employing a regime-switching model. Furthermore, the Wald test that results in a rejection of UFER hypothesis for the US-dollar/UK-pound series, provides evidence of unaccounted regime shifts in modelling the behaviour of the considered exchange rate series. This justifies further the use of a Markov regime-switching model in capturing better the shifts of the series.

A possible future direction of this work is related to currency option. A regime-switching Markov model may be introduced to capture the fluctuations of the underlying exchange rates in currency options. In fact, if a sudden movement happens and it cannot be forecasted, the option price is directly affected by this movement of the exchange rate. A regime-switching Markov model may be able to capture better this type of movements that are evidenced by changes amongst regimes.

References

1. Ang, A. and G. Bekaert (2002). "Regime switches in interest rates". *Journal of Business and Economic Statistics*, 20: 163-182.
2. Baillie, R. T. and P. C. McMahon. *The Foreign Exchange Market: Theory and Econometric Evidence.* (Cambridge University Press, Cambridge, 1989).
3. Bekaert, G. and R. J. Hodrick (1993). "On biases in the measurement of foreign exchange risk premiums". *Journal of International Money and Finance*, 12: 115-138.
4. Bilson, J. F. O. (1981). "The speculative efficiency hypothesis". *Journal of Business*, 54: 435-451.
5. Cornell, B. (1977). "Spot rates, forward rates and exchange market efficiency". *Journal of Financial Economics*, 5: 55-65.
6. Dahlquist, M. and G. Jonsson (1995). "The information in Swedish short-maturity forward rates". *European Economic Review*, 39: 1115-1132.
7. de Vries, C. G. (1994). "Stylised facts of nominal exchange rate returns". *Handbook of International Macreoeconomics*, (edited by F. van der Ploeg, Blackwell Ltd., Oxford (UK)-Cambridge (USA), 1994), 348-389.
8. Dickey, D. A. and W. A. Fuller (1979). "Distribution of the estimators for autoregressive time series with a unit root". *Journal of the American Statistical Association*, 74: 427-431.

9. Dickey, D. A. and W. A. Fuller (1981). "Likelihood ratio statistic for autoregressive time series with a unit root". *Econometrica*, 49(4): 1057-1072.
10. Diebold, F. X.. *Empirical Modelling of Exchange Rate Dynamics.* (Springer, Berlin, 1988).
11. Driffill, J. (1992). "Changes in regime and the term structure: A note". *Journal of Economic Dynamics and Control*, 16: 165-173.
12. Enders, W. *Applied Econometric Time Series.* (John Wiley & Sons, Inc., New York-Chichester-Brisbane-Toronto-Singapore, 1995).
13. Engel, C. M. (1996). "The forward discount anomaly and the risk premium: A survey of recent evidence". *Journal of Empirical Finance*, 3: 123-192.
14. Engel, C. M. and J. D. Hamilton (1990). "Long swings in the dollar: are they in the data and do markets know it?". *American Economic Review*, 80: 689-713.
15. Engle, R. E. and C. W. J. Granger (1987). "Cointegration and error correction: Representation, estimation and testing". *Econometrica*, 55: 251-276.
16. Evans, M. D. D. (1996). "Peso problems: Their theoretical and empirical implications". *Handbook of Statistic*, (edited by G. S. Maddala and C. R. Rao, Amsterdam: Elsevier Science, 1996), 613-646.
17. Evans, M. D. D. and K. K. Lewis (1994). "Do risk premia explain it all? Evidence from the term structure". *Journal of Monetary Economics*, 33: 285-318.
18. Evans, M. D. D. and K. K. Lewis (1995). "Do expected shifts in inflation affect estimates of the long run Fisher relation?". *Journal of Finance*, 50: 225-253.
19. Fama, E. (1984). "Forward and spot exchange rates". *Journal of Monetary Economics*, 14: 319-338.
20. Frenkel, J. (1980). "Exchange rates, prices and money: Lessons from the 1920s". *American Economic Review*, 70: 235-242.
21. Frenkel, J. and R. Meese (1987). "Are exchange rates excessively variable?". *NBER Macroeconomic Annual*, (edited by S. Fischer, MA, Cambridge, National Bureau of Economic Research, 1987), 117-162.
22. Froot, K. A. and J. A. Frenkel (1989). "Forward discount bias: Is it an exchange risk premium?". *Quarterly Journal of Economics*, 104: 139-161.
23. Gerlach, S. and F. Smets (1997). "The term structure of euro-rates: Some evidence in support of the EH". *Journal of International Money and Finance*, 16: 305-321.
24. Granger, C. W. J. and P. Newbold (1974). "Spurious regressions in econometrics". *Journal of Econometrics*, 2: 111-120.
25. Gray, S. F. (1996). "Modelling the conditional distribution of interest rates as a regime-switching process". *Journal of Financial Economics*, 42: 27-62.
26. Hakkio, C. S. and M. Rush (1989). "Market efficiency and cointegration: An application to the sterling and Deutsch mark exchange markets". *Journal of International Money and Finance*, 8: 75-88.
27. Hall, S. G., Psaradakis, Z. and M. Sola (1997). "Cointegration and changes in regime: The Japanese consuption function". *Journal of Applied Econometrics*, 12: 151-168.
28. Hamilton, J. D. (1988). "Rational expectations econometric analysis of changes in regime: An investigation of the term structure of interest rates". *Journal of Economic Dynamics and Control*, 12: 385-423.
29. Hamilton, J. D. (1989). "A new approach to the economic analysis of nonstationary time series and the business cycle". *Econometrica*, 57: 357-384.

30. Hardouvelis, G. A. (1994). "The term structure spread and future changes in long and short rates in the G-7 countries: Is there a puzzle?". *Journal of Monetary Economics*, 33: 255-283.
31. Isard P. *Exchange Rate Economics*. Cambridge Surveys of Economic literature, (MA: MIT Press, Cambridge, 1995).
32. Kim C. J. and C. R. Nelson. *State-Space Models with Regime Switching: Classical and Gibbs-Sampling Approaches with Applications*. (Cambridge University Press, Cambridge, 1995).
33. Levich, R. M. (1979). "On the efficiency of markets for foreign exchange". *International Economic Policy Theory and Evidence* (edited by Dornbusch, R., Frenkel, J., John Opkins Press) 246-267.
34. Levich, R. M. (1985). "Empirical studies of exchange rates: Price behaviour, rate determination and market efficiency". *Handbook of International Economics*, vol.2 (edited by R. W. Jones and P. B. Kenen, Amsterdam-North Holland, 1985), 979-1040.
35. Lewis, K. K. (1991). "Was there a peso problem in the U.S. term structure of interest rates: 1979-1982?". *International Economic Review*, 32: 159-173.
36. Meese, R. and K. J. Singleton (1982). "On unit roots and the empirical modelling of exchange rates". *Journal of Finance*, 37: 1029-1035.
37. Mussa, M. (1979). "Empirical regularities in the behaviour of exchange rates and theories of the foreign exchange market". *Carnegie-Rochester Conference on Public Policy*, 11: 9-57.
38. Paap, R. and H. K. van Dijk (2003). "Bayes estimates of Markov trends in possibly cointegrated time series: An application to U.S. consumption and income". *Journal of Business and Economic Statistic*, 21: 547-563.
39. Psaradakis, Z., Sola, M. and F. Spagnolo (2003). "On Markov error-correction models, with an application to stock prices and dividends". *Journal of Applied Econometrics*, 19: 69-88.
40. Sola, M. and J. Driffill (1994). "Testing the term structure of interest rates using a stationary vector autoregression with regime switching". *Journal of Economic Dynamics and Control*, 18: 601-628.
41. Spagnolo, F., Psazadakis, Z. and M. Sola (2004). "Testing the unbiased forward exchange rate hypothesis using a Markov-switching model and instrumental variables". *Journal of Applied Econometrics*, 20(3): 423-437.
42. Taylor, S. *Modelling Financial Time Series*. (John Wiley & Sons, Inc., Chichester, 1986).
43. Zivot, E. (2000). "Cointegration and forward and spot exchange rate regressions". *Journal of International Money and Finance*, 19: 785-812.

10

Early Warning Systems for Currency Crises: A Regime-Switching Approach

Abdul Abiad

International Monetary Fund
700 19th St. NW, Washington, DC 20431
USA
aabiad@imf.org

Summary. Previous early warning systems (EWS) for currency crises have relied on models that require a priori dating of crises. This paper proposes an alternative EWS, based on a Markov-switching model, which identifies and characterizes crisis periods endogenously; this also allows the model to utilize information contained in exchange rate dynamics. The model is estimated on data from 1972–1999 for the Asian crisis countries, taking a country-by-country approach. The model outperforms standard EWSs, both in signaling crises and reducing false alarms. Two lessons emerge. First, accounting for the dynamics of exchange rates is important. Second, different indicators matter for different countries, suggesting that the assumption of parameter constancy underlying panel estimates of EWSs may contribute to poor performance.

Key words: Currency crisis, early warning system, regime switching, Markov switching

10.1 Introduction

A succession of currency crisis episodes in the 1990s led to a proliferation of theoretical and empirical papers on the factors that brought about these crises. Several papers have also focused on the issue of anticipation–devising early warning systems that give policymakers and market participants warning that a crisis is likely to occur. Two approaches to constructing early warning systems have become standard: limited dependent variable probit/logit models and the indicators approach of Kaminsky, Lizondo and Reinhart [31], henceforth KLR. Berg et al. [4] assess the performance of these models, and find that they have outperformed alternative measures of vulnerability such as bond spreads and credit ratings. However, while these models are able to anticipate some crises, they also generate many false alarms.

There are several well-known methodological issues associated with the existing early warning models. Perhaps the most significant is that they require an *a priori* dating of crisis episodes before they can be estimated. The most common procedure for doing so is by taking changes in exchange rates, reserves and/or interest rates, choosing weights for each and combining them into an index of speculative pressure, specifying a sample-dependent threshold, and identifying crises based on whether or not the index exceeds the threshold. But as is evident from the survey of 26 recent empirical studies of currency crises in Section 10.2 below, this simple procedure has been applied in a multitude of ways, resulting in different periods being identified as crises[1].

The threshold procedure provides a set of crisis dates, but raises even more problems. First, the choice of the crisis-identification threshold is arbitrary. A selected sampling of thresholds used in the literature include the threshold of $1.5 \times \sigma$ (where σ is the sample standard deviation) used in Aziz et al. [2], $1.645 \times \sigma$ in Caramazza et al. [5], $1.75 \times \sigma$ in Kamin et al. [29], $2.5 \times \sigma$ in Edison [11], and $3 \times \sigma$ in KLR. Different choices of threshold will obviously result in different crisis dates and different estimated coefficients. Moreover, the threshold is sometimes treated as a free parameter and chosen so that the fit of the model is maximized (Kamin et al.[29]), or so that a set percentage, say 5 percent, of all observations are crises (Caramazza et. al.[5]).

Second, the sample-dependent nature of the threshold definition implies that future data can affect the identification of past crises. Thus one can observe cases of disappearing crises, as documented by Edison [11]. Since the threshold is defined in terms of the sample standard deviation, the occurrence of a new, relatively large crisis such as the Asian crisis results in previously identified crises no longer being identified as such. Edison notes that the threshold methodology identifies five crises in Malaysia using pre-1997 data, but these all disappear and only one crisis is identified (the 1997 crisis itself), when data up to 1999 are included in the sample.

Third, many of these studies make *ad hoc* adjustments to the binary crisis variable that may introduce artificial serial correlation. One common procedure is the use of "exclusion windows", which omits any crises identified by the threshold method if they follow a previous crisis within a certain window of time. As is the case with the threshold level, the width of the exclusion window is arbitrary, and has been chosen to be anywhere from one quarter (Eichengreen, Rose and Wyplosz [12]) to as long as 18 months (Aziz et al. [2]) and even 3 years (Frankel and Rose [16], using annual data). The motivation for using exclusion windows is to eliminate identifying speculative pressure episodes as new crises if they are just a continuation of a previous one. But in doing so, one eliminates any information the sample contains regarding

[1] For example, Kamin et al. [29] compare their identified crisis dates with those identified by KLR and find that only 61 percent of crisis dates were commonly identified.

crisis duration. More seriously, it introduces artificial serial correlation in the dependent variable that few studies account for. The estimated probit/logit models implicitly assume independence across observations t. But using an exclusion window means that $C_t = 1 \Rightarrow \Pr(C_{t+j} = 1) = 0$ for $j = 1, 2, \ldots, J$, where J is the width of the exclusion window[2].

Finally, information is lost when transforming a continuous variable into a binary variable. In particular, potentially useful information on the dynamics of the dependent variable is discarded. The critique regarding information loss can also be made regarding the treatment of the indicators in the KLR approach, where the explanatory variables themselves are transformed into binary signals.

Given these problems, is there an alternative approach? This paper proposes an EWS methodology, based on a Markov-switching model with time-varying transition probabilities, that can address these issues. First, the model does not require *a priori* dating of crisis episodes; instead, identification and characterization of crisis periods are part of the models output, estimated simultaneously with the crisis forecast probabilities in a maximum likelihood framework. One thus avoids the pitfalls associated with the threshold dating procedure described above. Additionally, by exploiting information in the dynamics of the dependent variable itself, the model is better able to send warning that a significant exchange rate adjustment is likely.

The assumptions that underlie a Markov-switching model are both concise and intuitive. The first assumption is that there are two states, tranquil periods and speculative attack periods. But we do not directly observe these states; that is, this binary "crisis" variable is *latent*. This brings us to our second assumption: there are directly observable variables whose behavior changes depending on the value of the crisis variable. Most obviously, the behavior of exchange rates is different during periods of speculative pressure than during tranquil periods[3]. In particular, we expect much greater exchange rate volatility as well as higher average depreciations during speculative attacks. Finally, we assume that given the current state–tranquil or crisis–there is a certain probability of staying in the same state, or moving to the other state. In our model, the probability of moving from the tranquil state to the crisis state depends on the strength or weakness of a countrys fundamentals.

Several studies have used Markov-switching models in developing theoretical models of speculative attacks. Jeanne and Masson [28] and Fratzscher [17]

[2] Another procedure that introduces artificial serial correlation, used by KLR and Berg and Pattillo [3] among others, is the practice of setting the dependent variable equal to one in the 24 months preceding crises identified using the threshold method. The rationale behind the procedure is to improve model fit for variables that exhibit abnormal behavior in the periods leading up to a crisis.

[3] One can substitute the speculative pressure index for the exchange rate if unsuccessful speculative attacks are also of interest; see Section 10.4 below.

develop currency crisis models with multiple equilibria and use a Markov-switching variable to model switches between these equilibria. In both cases, however, the probability of switching from one equilibrium to another is constant. In contrast, the model in this paper allows switches from the optimistic, no-attack equilibrium to the pessimistic, speculative attack equilibrium to be a function of various indicators.

Two other papers have used Markov-switching with time-varying probabilities to empirically model currency crises. Cerra and Saxena [7] use a Markov-switching model to look at the 1997 Indonesian crisis, and investigate whether the crisis was due to domestic factors, monsoonal factors, or pure contagion from neighboring countries. Their model differs from the one explored here, mainly because the only variable that affects the time-varying probability in their model is a measure of contagion, based on exchange market pressure in neighboring countries. Fundamentals in their model only affect the mean of the exchange rate. In contrast, our view is that domestic and external fundamentals affect the probability of a crisis occurring, and hence should enter into the time-varying probability equation rather than affecting only the level of the exchange rate.

The most closely related work is by Martinez-Peria [35], who also estimates a Markov-switching model with time-varying probabilities to model speculative attacks on the European Monetary System (EMS), using data from 1979 to 1993. That paper evaluated the ability of the Markov-switching model to identify crisis episodes[4], and assessed the degree to which five variables–domestic credit growth, the import-export ratio, the unemployment rate, the fiscal deficit and interest rates–determined crisis vulnerability in the EMS. We extend this work and focus primarily on the use of the model as an early-warning system. First of all, we begin by looking at a wider set of twenty-two early warning indicators. In addition to the standard macroeconomic indicators used in other early warning systems, we also explore indicators relating to the characteristics of capital flows, and those relating to financial sector soundness. Second, the predictive ability of the model is assessed both in-sample and out-of-sample. Finally, the Martinez-Peria study assumed that the parameters of the model are uniform across countries, and pooled the data to get parameter estimates. This is probably an innocuous assumption for the set of advanced economies in her study, which are broadly similar. But for developing countries, such an assumption might not hold. If a country is relatively more open or has less capital controls than other countries, for example, the coefficients on measures of external imbalance may be larger. In

[4] In this regard, the results are positive: Martinez-Peria finds that the Markov-switching model is able to identify all the crisis episodes identified by the methods used by Eichengreen, Rose and Wyplosz [12], but also identifies 25 additional crisis episodes. She then finds evidence in news reports and central bank releases that 21 of these 25 periods were indeed speculative attacks.

this paper, the model is estimated separately for each of the five Asian crisis countries (Indonesia, Korea, Malaysia, the Philippines and Thailand).

With the usual caveat that all early warning systems are far from perfect and serve only to synthesize information and supplement more in-depth country knowledge, the model does a good job of anticipating crises. It correctly anticipates two-thirds of crisis periods in sample, and just as important, sends much fewer false alarms than existing models. In the January 2000–July 2001 out-of-sample period, no warning signals are sent for three of the countries (Korea, Thailand and Indonesia), but vulnerabilities were signaled for Malaysia and the Philippines in mid-2001, mainly due to a decline in competitiveness and a slowdown in exports.

This chapter is organized as follows. Section 10.2 describes the Markov-switching model with time-varying probabilities in detail. The data used in the estimation is described in Section 10.3, while Section 10.4 presents the estimation results and a country-by-country analysis. Section 10.5 assesses the models predictive ability both in-sample and out-of-sample, and Section 10.6 concludes.

10.2 A Markov-switching approach to early warning systems

Regime-switching models have long been a tool available to empirical economists, with early work on these models going back to Quandt [36], Goldfeld and Quandt [19], and Hamilton [22]. Applications have only become common in the last decade, however, with the advent of greater computing power. Markov-switching models with constant transition probabilities have been applied to interest rates (Hamilton [20]), the behavior of GNP (Hamilton [21]), stock returns (Cecchetti, Lam and Mark [6]), and floating exchange rates (Engel and Hamilton [13])[5]. One serious limitation of the earlier Markov-switching models, however, was the restriction of constant transition probabilities. The baseline model was thus extended to allow for time-varying transition probabilities, by Lee [33] and Diebold, Weinbach and Lee [10] and used to model long swings in the dollar-pound rate, as well as by Filardo [15], [14] to analyze business cycle phases.

As mentioned in the introduction, there are two primary motivations for using Markov-switching model with time-varying probabilities in modeling speculative attacks. First, one can avoid the many ad hoc assumptions required in the standard models. Even if, as many of the studies claim, their results are robust to these ad hoc assumptions, we believe there is virtue in simplicity. Second, using exchange rates or the index of speculative pressure directly avoids the

[5] A comprehensive review of the applications of Markov-switching models in econometrics can be found in Kim and Nelson [32].

loss of information that results when these variables are transformed into a binary crisis dummy variable. In particular, exchange rate dynamics may itself be informative about the likelihood of a large speculative attack. A small increase in volatility (e.g., from a widening of an exchange rate band) or small devaluations in the span of a few months might foreshadow a coming currency crisis, but this information remains unutilized (and in fact is erased by the threshold dating process) in the standard approaches. As we will see below, even small changes in exchange rate behavior are utilized in a regime-switching framework as signs of increasing speculative pressure.

There are three disadvantages in using Markov-switching models. The first is computational; Markov-switching models with time-varying probabilities are still not part of the standard econometric software packages. But this drawback has become minor, as more researchers use the methodology and make their code available, and since software programs such as EViews now allow the creation of general log likelihood objects[6]. A second drawback is the difficulty in testing Markov-switching models against the null of no switching, as one encounters problems with unidentified nuisance parameters (the coefficient parameters in the transition probability matrix), as well as with a singular information matrix. Various tests have been suggested, including Davies [8], [9], Hansen [26], [27], Hamilton [24], Garcia [18] and Mariano and Gong [34] for testing a constant transition probability model against a null of no switching. For the time-varying transition probability case, one can do a sequential test: first, test the constant transition probability model against a null of no switching, and then test the time-varying transition probability model against a constant transition probability model. Note that testing the significance of individual coefficient estimates, as well as testing the overall model against a null of constant switching, can easily be done using standard t-statistics and likelihood ratio tests. The third drawback is that the likelihood surface can have several local maxima and is sometimes ill-behaved. The model may fail to converge when too many explanatory variables are included, and t-statistics may be sensitive to the choice of step size, since derivatives are calculated numerically. Thus, a judicious choice of start-up values and step size in the maximum likelihood estimation is important.

Model Specification and Estimation

The latent variable in the model follows a first-order, two-state Markov chain $\{s_t\}_{t=1}^{T}$, where $s_t = 1$ denotes a crisis state and $s_t = 0$ denotes a tranquil state. Although s_t is not directly observable, the behavior of our dependent variable y_t–which can be either the nominal exchange rate change or the speculative pressure index–is dependent on s_t as follows:

$$y_t | s_t \overset{\text{IID}}{\sim} N\left(\mu_{s_t}, \sigma_{s_t}^2\right) \tag{10.1}$$

[6] The EViews code and the dataset used for estimating the models in this paper are available from the author upon request.

so that both the mean and variance of y_t can shift with the regime[7]. The density of y_t conditional on s_t is then

$$f(y_t|s_t) = \frac{1}{\sqrt{2\pi}\sigma_{s_t}} \exp\left(\frac{-(y_t - \mu_{s_t})^2}{2\sigma_{s_t}^2}\right) \tag{10.2}$$

for $s_t = 0, 1$.

The latent regime-switching variable s_t evolves according to the transition probability matrix $\mathbf{P}_t$

$$\begin{array}{c} \\ \text{State 0} \\ \\ \\ \text{State 1} \\ \\ \end{array} \begin{array}{c} \text{State 0} \qquad\qquad\qquad\qquad\qquad \text{State 1} \\ \left[\begin{array}{cc} p_{00}^t & p_{01}^t = (1 - p_{00}^t) \\ \Pr(s_t = 0 \mid s_{t-1} = 0, x_{t-1}) & \Pr(s_t = 1 \mid s_{t-1} = 0, x_{t-1}) \\ = F(x_{t-1}\beta_0) & = 1 - F(x_{t-1}\beta_0) \\ & \\ p_{10}^t = (1 - p_{11}^t) & p_{11}^t \\ \Pr(s_t = 0 \mid s_{t-1} = 1, x_{t-1}) & \Pr(s_t = 1 \mid s_{t-1} = 1, x_{t-1}) \\ = F(x_{t-1}\beta_1) & = 1 - F(x_{t-1}\beta_1) \end{array}\right] \end{array} \tag{10.3}$$

where p_{ij}^t is the probability of going from state i in period $t-1$ to state j in period t, and F is a cumulative distribution function, most typically the logistic or the normal c.d.f. The elements of the $k \times 1$ vector x_{t-1} are the early warning indicators that can affect the transition probabilities.

One final quantity needed to complete the model is the start-up value $p_1^1 = \Pr(s_1 = 1)$, which gives the unconditional probability of being in state 1 at time 1. As Diebold, Weinbach and Lee [10] note, the treatment of this quantity depends on whether x_t is stationary or not. If x_t is stationary, then p_1^1 is simply the long-run probability that $s_1 = 1$, which in turn would be a function of (β_0, β_1). If x_t is nonstationary, then p_1^1 is an additional parameter that must be estimated. In practice, for a long enough time series this value has a negligible effect on the likelihood function, and whether one calculates it as a function of (β_0, β_1), estimates it as a separate parameter, or just sets it at a constant value makes little difference.

The estimation procedure we use is direct maximization of the likelihood, where the likelihood function is calculated using the iteration described in Hamilton [23, pp. 692-93]. Using information available up to time t, we can construct $\Pr(s_t = j \mid \Omega_t; \theta)$, the conditional (filtered) probability that the t-th observation was generated by regime j, for $j = 1, 2, \ldots, N$, where N is the number of states (in this paper, $N = 2$). Collect these conditional probabilities into an $(N \times 1)$ vector $\hat{\xi}_{t|t}$.

---

[7] An autoregressive process for y_t can be assumed as well, and the autoregressive parameters can switch from one regime to another, if desired. The assumption of normality can also be relaxed.

One can also form forecasts using the conditional (forecast) probability of being in regime j at time $t+1$, given information up to time t: $\Pr(s_{t+1} = j \mid \Omega_t; \theta)$, for $j = 1, 2, \ldots, N$. Collect these forecast probabilities in an $(N \times 1)$ vector $\hat{\xi}_{t+1|t}$. Lastly, let η_t denote the $(N \times 1)$ vector whose j-th element is the conditional density of y_t in equation (10.2). These filtered and forecast probabilities are calculated for each date t by iterating on the following equations:

$$\hat{\xi}_{t|t} = \frac{\hat{\xi}_{t|t-1} \circ \eta_t}{1' \left(\hat{\xi}_{t|t-1} \circ \eta_t \right)} \tag{10.4}$$

$$\hat{\xi}_{t+1|t} = \mathbf{P}'_{t+1} \hat{\xi}_{t|t} \tag{10.5}$$

where $\mathbf{P}_t$ is the $(N \times N)$ transition probability matrix going from period $t-1$ to period t, described in equation (10.3), and $\circ$ denotes element-by-element multiplication. Equation (10.4) calculates $\Pr(s_t = j \mid \Omega_t; \theta)$ as the ratio of the joint distribution $f(y_t, s_t = j \mid \Omega_t; \theta)$ to the marginal distribution $f(y_t \mid \Omega_t; \theta)$, the latter being obtained by summing the former over the states $1, 2, \ldots, N$. Equation (10.5) implies that once we have our best guess as to what state we are in today, we just pre-multiply by the transpose of the transition probability matrix $\mathbf{P}$ to obtain the forecast probabilities of being in various states in the next period.

Given an initial value for the parameters, θ, and for $\hat{\xi}_{1|0}$, which in our model is just $[1 - p_1^1, p_1^1]$, we can then iterate on (10.4) and (10.5) to obtain values of $\hat{\xi}_{t|t}$ and $\hat{\xi}_{t+1|t}$ for $t = 1, 2, \ldots, T$. The log likelihood function $L(\theta)$ can be computed from these as

$$L(\theta) = \sum_{t=1}^{T} \log f\left(y_t \mid X_t, Y_{t-1}; \theta\right) \tag{10.6}$$

where

$$f\left(y_t \mid X_t, Y_{t-1}; \theta\right) = 1' \left(\hat{\xi}_{t|t} \circ \eta_t \right) \tag{10.7}$$

One can then evaluate this at different values of θ to find the maximum likelihood estimate.

10.3 Data description and transformation

The model is estimated using monthly data from January 1972 to December 1999 for the five Asian crisis countries: Indonesia, Korea, Malaysia, the Philippines and Thailand. The dependent variable in our model is the month-to-month percentage change in the nominal exchange rate. Nothing precludes the use of the speculative pressure index as the dependent variable, if one is interested in unsuccessful speculative attacks as well. Another alternative to using the index of speculative pressure, which avoids the need to weigh the various components, is described in Abiad [1]. That paper adds reserve changes and interest rate changes as dependent variables, in addition to exchange rate changes, but rather than combining the three variables into a weighted average, the variables are stacked into a 3×1 vector whose distribution is dependent on the Markov-switching crisis variable s_t. The main finding is that adding reserve and interest rate changes to the model does not help identify any additional crisis episodes not already picked up by the univariate model based on exchange rates alone.

We explore a broad set of twenty-two early warning indicators, which are listed in Table 10.1. The indicators can be classified into three categories. The first group includes standard measures of *macroeconomic imbalance*. There are three measures of external imbalance: deviations of the real exchange rate from a Hodrick-Prescott trend[8], the current account balance relative to GDP, and the growth rate of exports. There are three measures of the adequacy of central bank reserves: the level and the growth rate of M2/reserves, and the growth rate of reserves. We also look at credit expansion, as measured by growth rate of real domestic credit. Two measures of real economic activity are used– the growth rate of industrial production and real GDP growth interpolated from quarterly data. Some crises have been preceded by the bursting of an asset market bubble, usually in the equities market or the property market, so we include the six-month change in the countrys stock market index as well. Finally, we include the real interest rate.

The second category of indicators relate to *capital flows*. The first indicator in this group is the 3-month LIBOR, which has been a primary determinant of the level of capital flows to emerging markets. A second indicator captures the idea that large capital inflows usually fuel a lending boom; one measure of this lending boom, first used by Sachs, Tornell and Velasco [37], is the growth in the ratio of bank assets to GDP. The three other indicators in this category focus on the composition of capital flows: the level of short-term debt to reserves, the stock of non-FDI investment (measured as a cumulation of flows) relative to GDP, and the ratio of cumulative portfolio inflows to total cumulative inflows.

The final category includes indicators of *financial fragility*. Kaminsky and Reinhart [30] noted that currency crises and banking crises tend to occur together, and that based on their sample of 20 countries over the period 1970–1995, problems in the banking sector typically precede a currency crisis. The first indicator of financial sector soundness we use is a rough measure of capital adequacy, the ratio of bank reserves to bank assets. A second indicator is central bank credit to banks, relative to total banking liabilities; an increase central bank credit may indicate financial weakness, if its purpose is to prop up or bail out weak banks. The ratio of bank deposits to M2 indicates the relative confidence that households and businesses have in the banking system,

[8] Alternative methods of detrending produce similar results.

Category	Concept		Measure
Macroeconomic indicators	External imbalance/real overvaluation	1	Deviations of real exchange rate from trend
		2	Current account balance/GDP
		3	Export growth rate
	Inadequacy of reserve cover	4	M2/Reserves, level
		5	M2/Reserves, growth rate
		6	Reserves growth rate
	Overexpansion of credit	7	Growth rate of real domestic credit, deflated by nominal GDP
	Slowdown in the real economy	8	Industrial Production, growth rate
		9	Real GDP, growth rate (interpolated from quarterly GDP)
	Asset price boom/bust	10	Stock market performance, growth rate
	Monetary tightening	11	Real interest rate
Capital flows indicators	Possible cause of reversal of flows	12	LIBOR
	Lending Boom	13	Bank assets/GDP, growth rate
	Short-term Debt	14	Short-term debt to reserves
	Composition of capital flows	15	Cumulative non-FDI flows/GDP
		16	Portfolio flows, share in stock of total capital flows
Financial fragility indicators	Capital adequacy	17	Bank reserves/Total bank assets
	Bailing out by the central bank	18	CB credit to banks/Total bank liabilities
	Confidence in banks	19	Bank deposits/M2, level
		20	Bank deposits/M2, growth rate
	Ability of banks to mobilize deposits	21	Loans/Deposits, level
		22	Loans/Deposits, growth rate

Table 10.1. Early Warning Indicators

with a low ratio indicating a lack of confidence; we include both the level and the growth rate of this ratio. Finally, we look at both the level and the growth rate of the loan-deposit ratio. A high and/or rising loans-to-deposits ratio may indicate increased banking system fragility, with an inadequate level of liquidity to respond to shocks. It should also be noted that one of the macroeconomic indicators, the real interest rate, is also frequently used as an indicator of financial sector soundness, as high real interest rates often lead to an increase in nonperforming loans[9].

Given the large number of indicators, a general-to-specific procedure was used to pare down the set and identify the final model. For each country, the model was run using each of the twenty-two early warning indicators, one at a time[10]. The coefficient estimates from these regressions can be found in Table 10.2. The coefficients on the indicators correspond to the parameter β_0 that enters into p_{00}^t, the probability of remaining in the tranquil state. All the variables are transformed such that an increase in the variable lowers the probability of remaining in the tranquil state, so that negative coefficient estimate is "correct".

Examining the results of Table 10.2 more closely, we see that the real overvaluation indicator is correctly signed and significant across all five countries. In fact, it is the only indicator that is uniformly correctly signed and significant. Four other variables–the level and growth rate of M2/reserves, the growth rate of real GDP, and the LIBOR–are correctly signed in all cases, but are only occasionally significant. All the other indicators have coefficient estimates that have correct signs and/or are significant for some countries, but not for others. Which brings us to another important point: indicator performance clearly varies widely from country to country. An assumption of parameter equality across countries, underlying EWS model estimates, which are based on a panel of countries, may lead to incorrect results, and may contribute to poor predictive performance.

The set of indicators for each country was then narrowed based on which coefficient estimates were correctly signed. Desire to monitor a wider set of indicators suggested against eliminating correctly signed coefficients whose

---

[9] Nonperforming loan ratios were also considered but were unavailable for most countries for a long enough period. Evidence for the Philippines, however, indicates that NPL ratios are a lagging rather than a leading indicator; the NPL ratio actually declined from about 8 percent in 1990 to about 4 percent just before the Asian crisis, and only started increasing substantially in late 1997 and in 1998.

[10] The hill-climbing method using in maximum likelihood estimation will converge very slowly if the various indicators are of very different magnitudes, since a step-size that is fine enough for the small variables will move very slowly for the large ones. To aid in estimation we rescale each variable to be zero mean and unit variance. An alternative transformation, used by KLR and Berg and Pattillo, is to use percentiles.

Indicator	Indonesia Coeff	Indonesia t-stat	Korea Coeff	Korea t-stat	Malaysia Coeff	Malaysia t-stat	Philippines Coeff	Philippines t-stat	Thailand Coeff	Thailand t-stat
Deviations of real exchange rate from trend	-0.65	-2.84	-0.30	-2.02	-0.35	-2.55	-0.30	-2.28	-0.26	-2.17
Current account balance/GDP	-0.36	-0.89	-0.97	-1.42	-0.12	-0.43	-0.12	-0.90	0.22	1.43
Export growth rate	-0.36	-0.83	-0.28	-1.04	0.11	1.10	-0.29	-2.09	0.02	0.15
M2/Reserves, level	-0.18	-1.29	-0.13	-1.37	-0.04	-0.33	-0.49	-3.62	-0.26	-1.91
M2/Reserves, growth rate	-0.16	-1.27	-0.14	-1.58	-0.09	-0.64	-0.22	-1.61	-0.28	-2.09
Reserves growth rate	-0.37	-1.52	0.03	0.23	0.04	0.29	-0.28	-1.45	-0.45	-2.13
Growth rate of real domestic credit, deflated by nominal GDP	0.04	0.26	-0.18	-1.07	-0.32	-2.16	0.47	1.91	-0.15	-0.81
Industrial Production, growth rate	-0.20	-1.29	-0.28	-1.39	0.00	-0.03	-0.07	-0.52	1.42	1.24
Real GDP, growth rate (interpolated from quarterly GDP)	-0.43	-1.05	-0.28	-1.22	-0.19	-1.20	-0.29	-2.39	-0.73	-2.69
Stock market performance, growth rate	-0.12	-0.06	-3.14	-1.27	-0.24	-1.01	0.32	1.83	-0.13	-0.89
Real interest rate	-0.17	-0.57	0.12	0.87	-0.15	-1.08	0.11	1.06	-0.39	-2.21
LIBOR	-0.11	-0.50	-0.01	-0.07	-0.27	-2.32	-0.56	-3.97	-0.14	-1.04
Bank assets/GDP, growth rate	0.17	0.92	-0.06	-0.35	-0.26	-2.13	0.27	2.24	0.02	0.10
Short-term debt to reserves	0.19	0.66	-0.01	-0.01	-0.74	-2.92	-1.01	-3.95	-0.25	-0.91
Non-FDI capital flows; stock relative to GDP	0.26	0.81	-0.01	-0.02	-0.01	-0.07	-0.70	-2.41	-0.43	-1.65
Portfolio flows, share in stock of total capital flows	-0.22	-1.24	-0.40	-2.39	0.10	0.55	-0.29	-1.86	-0.35	-2.00
Bank reserves/Total bank assets	-0.05	-0.18	-0.25	-1.34	0.07	0.44	0.21	1.64	-0.04	-0.28
CB credit to banks/Total bank liabilities	-0.24	-1.76	0.02	0.18	0.02	0.07	-0.47	-3.62	0.04	0.26
Bank deposits/M2, level	0.00	0.00	0.10	0.79	-0.16	-1.14	0.00	-0.03	0.21	1.26
Bank deposits/M2, growth rate	-0.23	-0.58	0.13	1.01	-0.04	-0.32	-0.10	-1.01	0.13	0.65
Loans/Deposits, level	0.14	0.33	-0.13	-0.41	0.07	0.53	-0.19	-1.58	0.02	0.11
Loans/Deposits, growth rate	0.03	0.13	-0.13	-0.87	-0.08	-0.57	0.17	1.81	0.46	1.61

Table 10.2. Coefficient Estimates for the 22 EWIs (Bivariate Regressions)

t-statistics were not significant at the 5 percent level[11]. The moderate correlation among the early warning indicators also suggested that the t-statistics may be misleading. In addition, a likelihood-ratio test of the joint significance of the explanatory variables showed them to be significant.

How high must forecast probabilities rise to be warranted as significant? It has been standard practice in the early-warning systems literature to map a models forecast probability into a binary "alarm signal" by determining some cutoff probability, and letting the signal equal 1 if the forecast probability rose above this threshold, and 0 otherwise[12]. An assessment of predictive ability is then conducted by computing the number of crises the model signal correctly calls (by sending an alarm within a particular window, usually 24 months before the actual crisis occurrence), and the number of false alarms the model sends.

In this context, it should be noted that forecast probabilities between competing early warning systems are not directly comparable; in particular, one should adjust for the time horizon the model is using. Most of the early warning systems in the literature focus on relatively long-horizon forecasting, with horizons of 12 or 24 months being the norm. The regime-switching model we use here, on the other hand, estimates one-month ahead forecasts. To make forecast probabilities from different models comparable, the forecast horizons must be matched, and the most straightforward way to do this would be to transform the short-horizon forecast into a long-horizon equivalent, using:

$$\begin{aligned} \Pr(\text{crisis over next } n \text{ months}) &= 1 - \Pr(\text{no crisis over next } n \text{ months}) \\ &= 1 - (\Pr(\text{no crisis over next 1 month}))^n \\ &= 1 - (1 - \Pr(\text{crisis over next 1 month}))^n \end{aligned} \tag{10.8}$$

Of course, this transformation is made under the assumption that the fundamentals that determine the crisis probability neither worsen nor improve. If the former, then the n-month crisis probability will be higher; if the latter, then the crisis probability will be lower[13]. As an example, a 10 percent

[11] The more conventional procedure of keeping only correctly signed and statistically significant variables results in only one indicator (real overvaluation) remaining for Indonesia, Korea and Malaysia, and only two indicators remaining for the Philippines (real overvaluation and M2/reserves) and Thailand (real overvaluation and real GDP growth).

[12] Although originally applied to individual indicators by Kaminsky, Lizondo and Reinhart [31], this methodology has since been used on the overall model by several studies, most notably Berg et al. [3], to evaluate the model's performance, as well as to compare competing early warning systems.

[13] Alternatively, one can construct projected time paths for the early warning indicators, and use these to calculate an n-month crisis probability. The accuracy will, of course, depend on the reliability of the projected time paths.

probability of a crisis over one month would be equivalent, *ceteris paribus,* to a three-month crisis probability of $1-(0.90)^3 = 27$ percent, and a one-year crisis probability of $1-(0.90)^{12} = 72$ percent.

10.4 Estimation results

The final model estimates for the five countries can be found in Table 10.3. For all five countries, State 0 is identified as a low-mean, low-volatility regime while State 1 is a high-mean, high-volatility regime. Average volatility, as measured by the standard deviation, is very low in the tranquil state–less than 1 percent per month in all five countries–while average volatility during crisis periods is quite large, with the highest crisis volatilities estimated for Indonesia, at 29 percent per month. In fact, volatility seems to be the primary distinguishing characteristic between tranquil and crisis periods, as σ_1 is significantly different from σ_0 in all cases. The average depreciation in tranquil periods is effectively zero (less than a quarter percent per month) in all countries, while in crisis periods it ranges from 2.1 percent per month in Thailand, to 12.6 percent per month in Indonesia, but the standard errors are large enough so that one cannot reject equality of μ_1 and μ_0. The coefficients on the indicators in the time-varying probabilities are all correctly signed, but as noted earlier, they are insignificant in most cases. This might be due to correlations among the indicator variables; in fact, likelihood-ratio tests for the joint significance of the indicators are significant for all countries except Malaysia, where the test of joint significance is marginally insignificant, with a p-value of 0.16. We now turn to a country-by-country analysis.

10.4.1 Indonesia

In the estimated model for Indonesia (Table 10.3), six indicators are used–real overvaluation, export growth, the level of M2/reserves, reserve growth, central bank credit to the banking sector, and growth of the M2/deposits ratio. There are five speculative pressure episodes in Indonesia in our 1972-1999 sample (Table 10.4)–a devaluation of 50 percent in November 1978, currency volatility in late 1982 that culminated in a 38 percent devaluation in April 1983, moderate volatility and a 5 percent depreciation in mid-1984, a 44 percent devaluation in September 1986, and the Asian crisis which began in July 1997 with a 6 percent decline in the rupiah. Figure 10.1 plots these crisis dates, along with 12-month forecast probabilities and alarm signals based on a 50-percent cutoff. Alarm signals are sent at least once in the 12 months preceding four of the five crisis episodes, with the only uncalled crisis being the smallest one, the 5 percent depreciation in mid-1984. However, the Asian crisis was not well-signaled for Indonesia; an alarm was generated only in one month (October 1996), and reflected increased currency volatility during that period. The forecast probabilities do increase steadily, to 45 percent in June 1997, but stay below the signaling threshold of 50 percent.

Indicator	Indonesia		Korea		Malaysia		Philippines		Thailand	
	Coeff	t-stat	Coeff	t-stat	Coeff	t-stat	Coeff	t-stat	Coeff	t-stat
Mean, State 0	0.2	11.13	0.15	3.09	0.01	0.14	0.05	2.44	0	0.09
Mean, State 1	12.59	1.32	2.83	0.73	4.41	1.96	2.62	2.27	2.1	1.35
Sigma, State 0	0.3	27.43	0.73	31.76	0.83	21.34	0.21	21.36	0.46	27.61
Sigma, State 1	29.54	8.55	12.18	7.99	4.15	4.65	6.36	13.39	8.4	10.01
Constant (beta0)	2.52	4.42	17.14	0.46	1.76	6.18	1.92	5.7	2.9	1.64
Deviations of real exchange rate from trend	-0.6	-1	-1.13	-0.41	-0.21	-0.93	-0.47	-1.91	-0.1	-0.27
Current account balance/GDP			-4.39	-0.24						
Export growth rate	-0.05	-0.08					-0.14	-0.32		
M2/Reserves, level	-0.16	-0.45	-1.48	-0.12			-0.74	-3.23	-0.58	-0.28
M2/Reserves, growth rate							-0.12	-0.6		
Reserves growth rate	-0.08	-0.18							-0.62	-0.6
Growth rate of real domestic credit, deflated by nominal GDP					-0.2	-0.82				
Industrial Production, growth rate			-3.74	-0.51			-0.31	-1.14		
Real GDP, growth rate (interpolated from quarterly GDP)					-0.29	-0.93			-0.21	-0.12
Stock market performance, growth rate			-7.25	-0.69						
Real interest rate					-0.04	-0.15			-0.07	-0.04
LIBOR					-0.14	-0.69				
Non-FDI capital flows; stock relative to GDP									-1.04	-0.32
Portfolio flows, share in stock of total capital flows			-2.35	-0.36						
CB credit to banks/Total bank liabilities	-0.34	-0.86								
Bank deposits/M2, level					-0.26	-0.47				
Bank deposits/M2, growth rate	-0.29	-0.51					-0.03	-0.13		
Constant (beta1)	0.71	2.44	0.94	3.4	-0.02	-0.03	0.65	3.84	0.82	3.02
Number of observations	336		276		288		216		276	
LR Test Statistic for Joint Significance of Indicators	11.68		17.61		9.17		25.92		21.07	
p-value	0.07		0.01		0.16		0		0	

Table 10.3. Final Model Estimates

10.4.2 Korea

The indicators that enter the final model for Korea are real overvaluation, the current account to GDP ratio, the level of M2/reserves, industrial production growth, stock market performance, and the share of portfolio flows in total capital flows (Table 10.3). Three crisis periods are included in the sample (Table 10.5)–a 20 percent devaluation in January 1980, a depreciation of 7 percent in September-November of the same year (which was likely a continuation of earlier speculative pressure), and the Asian crisis (Figure 10.2).[14] Interestingly, the model already identifies March 1997, when the won depreciated by 4 percent, as a period of speculative pressure. In terms of predicting these episodes, the model does not anticipate the January 1980 depreciation; however, after the initial devaluation it continues to send signals in anticipation of further speculative pressure, which did occur later that year. With regard to the Asian crisis, the Korean model illustrates the gains from letting an EWS model use information available in the exchange rate behavior. There was already a moderate increase in the wons volatility even before the Asian crisis, beginning as early as the middle of 1996, when the won depreciated by 3 percent. As a result of this increased volatility–and combined with Korea's weakening external position, a decline in the stock market and the high share of portfolio flows–the model begins signaling in February 1997.

10.4.3 Malaysia

The final model for Malaysia contains six indicators–real overvaluation, domestic credit growth, real GDP growth, the real interest rate, the LIBOR, and the ratio of M2 to deposits. Relative to the four other countries, Malaysia's exchange rate regime was much less of a peg and more of a dirty float. Thus, unlike other countries which experienced rarer but sharper devaluations, the speculative pressure episodes in Malaysia are protracted periods characterized by increased volatility and a slow deterioration of the exchange rate. Thus, instead of identifying the individual spikes, we group them into four periods which are described in Table 10.6.

The model is able to anticipate three of Malaysia's four speculative pressure periods (Figure 10.3). Analyzing the individual indicators that enter the model, one finds that the rise in world interest rates contributed to the speculative pressure that occurred in the late 1970s and early 1980s; overvaluation, high real interest rates and a slowdown in real growth contributed to the 1985-1986 depreciation; and overvaluation and a domestic credit boom increased vulnerability in the run-up to the Asian crisis.

[14] Current account data for Korea only begins in 1977; hence two earlier devaluations of 14 percent in June 1971 and 21 percent in December 1974 are not covered. The small number of crises and the closeness of the fit in Figure 2 raise the concern that the model may be overfitting the data in the Korean case.

Episodes Identified	Description	Signaled/Anticipated by Model?
November 1978	Devaluation of 50 percent	Yes, from February-September 1978
October 1982-April 1983	Currency volatility; devaluation of 38 percent in April 1983	Yes, from August 1982 to September 1983 (sporadic signals)
August-September 1984	Devaluation of 4 1/2 percent over three months	No
September 1986	Devaluation of 44 percent	Yes, from February 1985 to June 1986
July 1997 onwards	Asian Crisis	Yes, but only one month (October 1996) due to increased volatility in the rupiah; probabilities rise from 22 percent in November 1996 to 45 percent in June 1997, but do not cross 50 percent signal threshold

Table 10.4. Speculative Pressure Episodes and Alarm Signals in Indonesia

Episodes Identified	Description	Signaled/Anticipated by Model?
January 1980	Devaluation of 20 percent	No
October 1980	Depreciation of 7 percent over three months	Yes, from January-May 1980 (i.e., more volatility was expected after January depreciation)
March 1997; and October 1997 onwards	Asian Crisis (October 1997 onwards), but model also detects increased volatility in early 1997 and already identifies March depreciation (4 percent) as speculative pressure period	Yes, but only from February 1997 onwards

Table 10.5. Speculative Pressure Episodes and Alarm Signals in Korea

10.4.4 The Philippines

The six indicators that enter into the Philippine model are real overvaluation, export growth, both the level and the growth rate of M2/reserves, industrial production growth, and the growth rate of deposits/M2. The model was estimated from 1982 onwards, since data on industrial production growth is unavailable before 1982. The model identifies three protracted periods of spec-

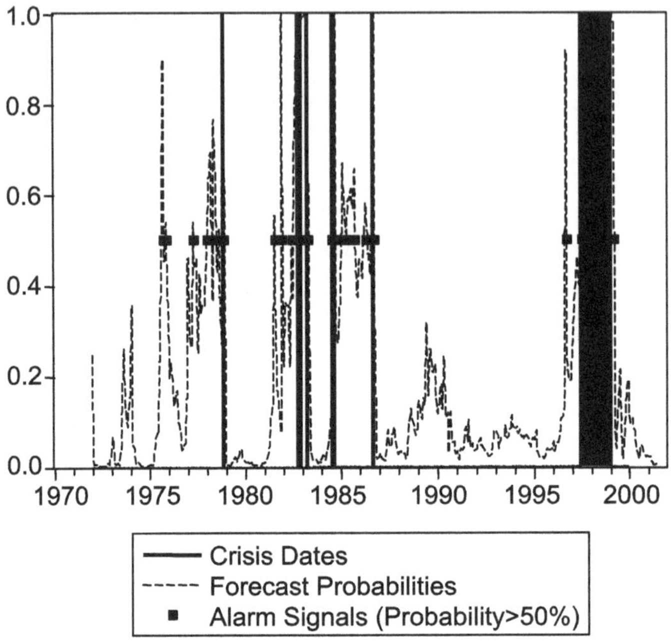

Fig. 10.1. Indonesia: Crisis Dates, Forecast Probabilities and Alarm Signals

Episodes Identified	Description	Signaled/Anticipated by Model?
October 1978-June 1982	Volatility; 12 percent depreciation over the period	Yes, from November 1977
January 1985-March 1986	Depreciation of 15 percent	Yes, from August 1984
December 1992-January 1994	Depreciation of 9 percent	No
July 1997 onwards	Asian Crisis	Yes, from January 1997

Table 10.6. Speculative Pressure Episodes and Alarm Signals in Malaysia

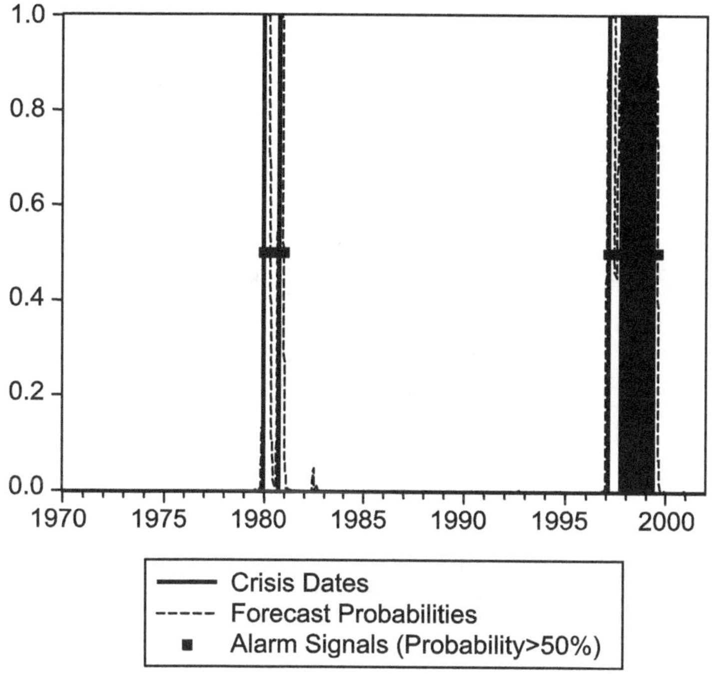

Fig. 10.2. Korea: Crisis Dates, Forecast Probabilities and Alarm Signals

ulative pressure (Table 10.7).[15] The first is from August 1982 to April 1986, when a financial crisis and political turmoil resulted in high exchange rate volatility and a 140 percent depreciation over the period. The second is from June 1988, a period of moderate volatility where the peso depreciated by 34 percent. The third period is the Asian crisis, which began for the Philippines with a 10 percent depreciation in July 1997.

The model for the Philippines is able to anticipate these three crisis periods (Figure 10.4). Analyzing the individual indicators, one finds that different factors were behind each crisis. A slowdown in both exports and industrial production played some role in triggering the crisis in the early 1980s, but a rise in both the level and growth rate of M2/reserves, as well as a sharp fall in the deposits/M2 ratio (an indicator of the banking crisis that occurred), played a role in prolonging the crisis. Reserve adequacy, as measured by both the level and growth rate of M2/reserves, also increased vulnerability in the

[15] Note that the period in the early to mid 1990s, when the peso was allowed to float more freely, was omitted so as not to be identified as a crisis period.

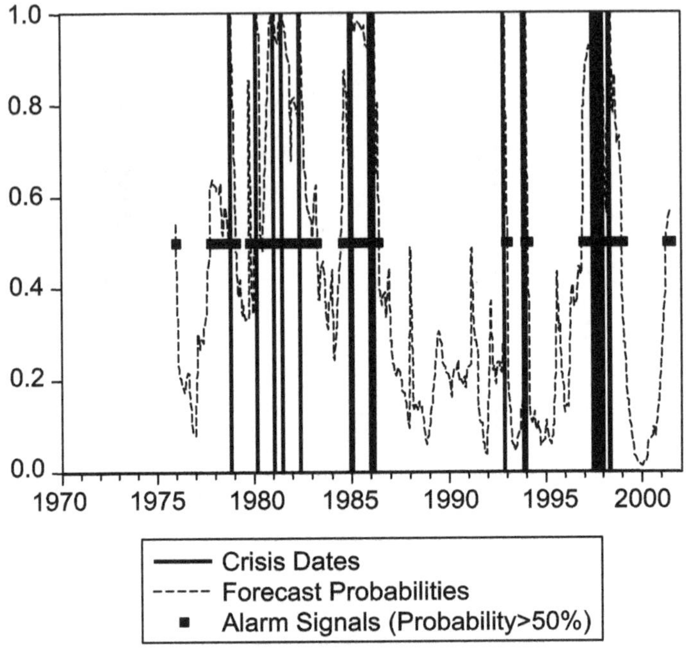

Fig. 10.3. Malaysia: Crisis Dates, Forecast Probabilities and Alarm Signals

Episodes Identified	Description	Signaled/Anticipated by Model?
August 1982-April 1986	Depreciation of 140 percent; high volatility (s.d. 7 percent)	Yes, from January 1982
June 1988-November 1990	Depreciation of 34 percent; moderate volatility (s.d. 2 percent)	Yes, from December 1987
July 1997 onwards	Asian Crisis	Yes, from May 1996

Table 10.7. Speculative Pressure Episodes and Alarm Signals in the Philippines

late 1980s, which culminated in 17 percent depreciation in the latter half of 1990, during the Gulf War. Finally, weakening competitiveness that began in late 1996–resulting from the appreciation of the yen against the dollar, to which the peso was pegged–increased the Philippines vulnerability, and resulted in the depreciation of the peso following the float of the Thai baht in July 1997.

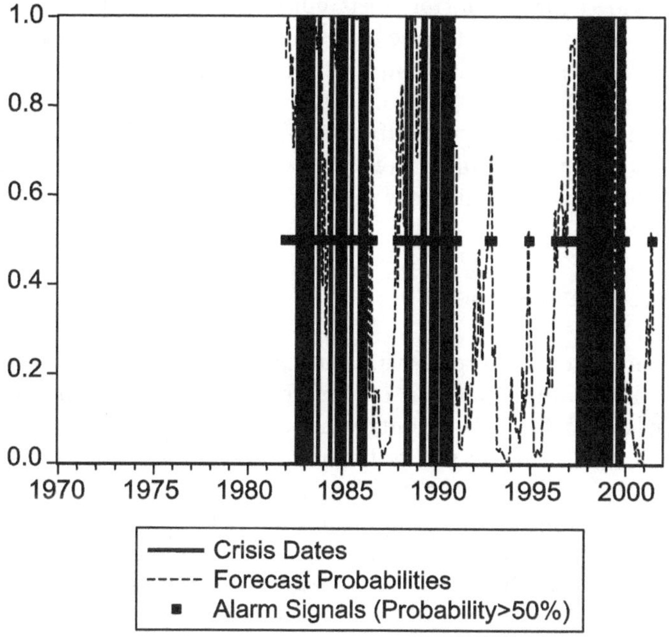

Fig. 10.4. Philippines: Crisis Dates, Forecast Probabilities and Alarm Signals

10.4.5 Thailand

The final model for Thailand includes the following indicators: real overvaluation, the level of M2/reserves, reserve growth, real GDP growth, the real interest rate, and the share of non-FDI capital flows in total flows. There are three crisis periods in the sample, a 10 percent depreciation in 1981, a 19 percent depreciation between November 1984-December 1985, and the Asian crisis which began in Thailand in July 1997 (Table 10.8).

Episodes Identified	Description	Signaled/Anticipated by Model?
July 1981	Devaluation of 10 percent	Yes, but only two months ahead (May 1981)
November 1984–December 1985	Depreciation of 19 percent	Yes, in December 1983–January 1984 and from July 1984
July 1997 onwards	Asian Crisis	Yes, from December 1996

Table 10.8. Speculative Pressure Episodes and Alarm Signals in Thailand

All three crisis periods are anticipated, as can be seen in Figure 10.5. However, signals are sent only two months prior to the July 1981 devaluation. Better warning is provided for the latter crisis episodes. Real overvaluation seems to have played some role in all three crises. Reserve inadequacy, as measured by the level of M2/reserves, also played a role in the 1980s episodes, but not in the Asian crisis. Based on the model, three factors seem to have played a role in increasing Thailands vulnerability to crisis in 1997–a loss of external competitiveness, a slowdown in the real economy, and an increasing proportion of non-FDI flows in total capital flows.

10.5 Forecast assessment

We now perform a more rigorous evaluation of the predictive performance of the model, both in-sample and out-of-sample. Table 10.9 contains in sample goodness-of-fit tables for each country model, as well as overall for all five countries. Each 2×2 matrix shows the number of correctly called tranquil and crisis periods, as well as the number of false alarms and the number of missed signals. We summarize this information further in Table 10.10, which also provides goodness-of-fit measures for five other models evaluated in Berg and Pattillo [3], henceforth BP. The comparison is only meant to be indicative, as the Markov-switching model differs from the five other models in several important ways, beyond just the differences in model specification. First, the identified crisis dates are different. Second, the forecast horizons are not the same–the models reviewed by BP all use a forecast horizon of 24 months, as opposed to the 12-month forecast horizon used here. Third, the data underlying the estimates, and the transformations applied to them, are similar but not identical. Finally, the models in this paper are estimated country-by-country, whereas all five BP models were estimated on a panel of countries, a point we return to below.

We see that Markov-switching model correctly calls 81 percent of observations. This is slightly lower than the 82-85 percent performance of the standard models in BP, when they use a 50 percent cutoff probability. However, the high predictive performance in those models is driven mostly by their ability to call tranquil periods correctly; they correctly classify 98-100 percent of tranquil periods, as opposed to 89 percent in the Markov-switching model. But in terms of correctly called crises, the Markov-switching model performs much better, calling 65 percent of pre-crisis periods correctly–that is, sending signals in 65 percent of the months where a crisis ensued within a years time. The standard models, in contrast, only call 7-19 percent of pre-crisis months correctly. The poorer performance is probably due in part to the longer 24-month forecast horizon they aim for, and also because the 50 percent signaling threshold is too high for those models. BP also report goodness-of-fit for the standard models when the cut-off is lowered to 25 percent, which we replicate

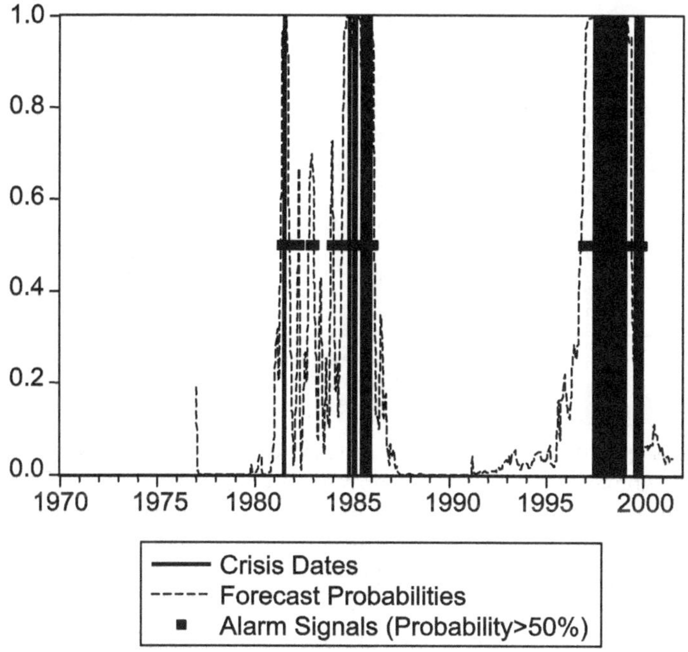

Fig. 10.5. Thailand: Crisis Dates, Forecast Probabilities and Alarm Signals

the bottom half of Table 10.10. The lower signaling threshold increases the number of correctly called pre-crisis periods–now the models correctly send signals in 41-48 percent of pre-crisis periods–but this comes at the expense of a much higher fraction of false alarms. With the lower threshold, false alarms account for 57-65 percent of total alarms, i.e., almost two out of every three signals are false alarms.

Although the Markov-switching specification probably accounts for part of the improved performance, it is also possible that a substantial portion of the improved performance is due to the fact that the standard models estimate the data using a panel of countries, and assuming that the coefficients are uniform across countries. As we saw in Section 10.5, indicators that matter for crises in one country may not even be pointing in the right direction during crises in another country.

What are the out-of-sample predictions of the country Markov-switching models? The models were estimated using data up to the end of 1999. An attempt to estimate the model up to end-1996, to see whether the Asian crisis was forecastable using the model, was not possible in this case, mainly because

		Crisis within 12 mos.	No crisis within 12 mos.
Overall	Signal	270	102
	Non-signal	147	812
Indonesia	Signal	38	18
	Non-signal	47	221
Korea	Signal	26	4
	Non-signal	29	205
Malaysia	Signal	77	29
	Non-signal	48	121
Philippines	Signal	83	40
	Non-signal	2	79
Thailand	Signal	46	11
	Non-signal	21	186

Table 10.9. Forecast Assessment

the model was estimated country-by-country; eliminating the Asian crisis not only removes the most informative episode in the sample, but also results in overfitting and/or nonconvergence of the maximum likelihood algorithm. Hence the only alternative is to look at model forecasts beyond the end of 1999. Admittedly, the hold-out sample from January 2000–July 2001 is relatively small, and moreover, none of the five countries had a crisis during this period. But it is still an informative exercise to see what kinds of probabilities and signals the country models send.

The forecast probabilities and alarm signals for the out-of-sample period of January 2000–July 2001 can be seen in Figures 10.1–10.5. For three of the countries–Indonesia, Korea and Thailand–no alarm signals are sent during the period. There was still a moderate probability (about 20 percent) of a crisis in Indonesia through much of 2000, but vulnerabilities (at least those measured by the indicators in the model) have dropped since then. Thailand has shown lower susceptibility to a crisis, and Korea even less so.

In contrast, the models for Malaysia and the Philippines did signal some vulnerability in the out-of-sample period, although only for a few months. Crisis probabilities in Malaysia were actually dropping toward the end of the estimation period (1999) and were low through most of 2000, but started increasing in the last quarter of 2000 and accelerating in 2001 up until July 2001, the last available data point. In fact, probabilities were high enough that the model began sending signals in May 2001. What was driving this increase in vulnerability? An analysis of the indicators entering the Malaysian model identifies several weaknesses. First, there was a steady decline in competitiveness, as measured by the real exchange rate, through 2000 and 2001. [16] Second, there

[16] This was also evident in another indicator that does not enter into the Malaysian model, export growth.

		From Berg and Pattillo [3]:				
	Markov-Switching Model	KLR Original	KLR Augmented	BP Indicators Probit	BP Linear Probit	BP Piecewise Linear Probit
Goodness-of-fit (cut-off probability of 50%)						
Percent of observations correctly called	81	82	83	85	84	85
Percent of pre-crisis periods correctly called[a]	65	9	9	16	7	19
Percent of tranquil periods correctly called[b]	89	98	99	99	100	98
False alarms as percent of total alarms[c]	27	44	30	29	11	34
Goodness-of-fit (cut-off probability of 50%)						
Percent of observations correctly called		77	75	81	78	80
Percent of pre-crisis periods correctly called[a]		41	46	44	48	47
Percent of tranquil periods correctly called[b]		85	81	89	84	87
False alarms as percent of total alarms[c]		63	65	57	63	59

[a] A pre-crisis period is correctly called when the estimated probability of crisis is above the cut-off probability and the crisis ensues within 12 months (Abiad), or within 24 months (other models).

[b] A tranquil period is correctly called when the estimated probability of crisis is below the cut-off probability and no crisis ensues within 12 months (Abiad), or within 24 months (other models).

[c] A false alarm is an observation with an estimated probability of crisis above the cut-off (an alarm) not followed by a crisis within 12 months (Abiad), or within 24 months (other models).

Table 10.10. Measures of Predictive Power

was a slowdown in the real economy. And third, there was a sharp rise in real domestic credit growth.

The Philippines also showed some weaknesses in the out-of-sample period, according to the model. Crisis probabilities were actually low in 2000, but the model starts indicating moderate vulnerabilities beginning the second quarter of 2001. A spike in the crisis probability led to a signal in June 2001, but probabilities decreased in July, the last data point. There was one primary factor behind the increased vulnerability in the Philippines: a weakened external position, seen most clearly in rapidly contracting exports. Over the January 2000–July 2001 out-of-sample period, then, signals were sent for only 4 out of 95 months: May-July 2001 for Malaysia, and June 2001 for the Philippines, and these reflected vulnerabilities due to external weaknesses present in these two countries at that time.

10.6 Conclusions

There is a general consensus among economists that early warning systems, no matter how sophisticated, will not be able to forecast crises with a high degree of accuracy. Even economists who construct such models are aware of this, and see these models as no more than useful supplements to more informed country analyses, and as a means of summarizing information in an unbiased, objective manner. Nevertheless, increased emphasis on crisis prevention (as opposed to crisis resolution) means that policymakers need to utilize all the tools available for assessing countries vulnerabilities, and to improve these tools when possible. This paper hopes to assist in this effort, first by surveying the recent empirical literature on currency crises, and by analyzing an alternative EWS approach that addresses some of the shortcomings of existing models.

The survey of 30 selected empirical studies written since 1998 is meant to increase awareness of the various econometric approaches to early warning systems that have been developed, so that practitioners have at their disposal a larger set of tools in assessing vulnerability. Many of the proposed approaches look promising, and virtually all report some improvement over the standard probit/logit and indicators models. However, many of the studies do not perform rigorous evaluations of performance. Adoption of standard evaluation procedures–including goodness-of-fit tables and measures, accuracy scores, and out-of-sample testing–will help potential users gauge how useful these models really are. Furthermore, it is difficult to assess relative performance across models, given differences in the datasets used and in the sample of countries studied. A true "horse race" among competing models–where each specification is estimated using the same data and sample of countries–will help resolve this issue.

But even in-sample and out-of-sample tests are only indicative; the true test of these models is in operationalizing them. In this regard, an additional measure of a models usefulness is simplicity of application. Early warning systems should be easy to replicate and estimate. That is, there should be minimal reliance on ad hoc assumptions, the data should come from published sources, and one should ideally be able to estimate the model using standard software packages or with programming code provided by the authors. If these conditions are satisfied, then it should be possible to monitor these models in real-time at low cost.

In addition to surveying the recent literature on early warning systems, this paper also contributes to it by suggesting an alternative EWS approach based on a Markov-switching model with time-varying transition probabilities. The model does an adequate job of anticipating crises. It correctly anticipates two-thirds of crisis periods in sample (compared to about 50 percent for the standard models), and just as important, sends a much smaller proportion of false alarms. In the January 2000–July 2001 out-of-sample period, no warning signals are sent for three of the five countries studied (Korea, Thailand and Indonesia), but vulnerabilities were signaled for Malaysia and the Philippines in mid-2001, mainly due to a decline in competitiveness and a slowdown in exports.

Beyond the performance of the Markov-switching model itself, there are some lessons that apply to the construction of early warning systems in general. First, accounting for dynamics is important. There is useful information in both the level and the volatility of the exchange rate itself that existing models have ignored. More specifically, some crises have been preceded by a series of smaller depreciations, by a widening of an exchange rate band, and/or an increase in the volatility of exchange rates, and this has not been utilized in existing models. Second, although there are some indicators which are common across countries in their predictive ability (with the real exchange rate being the most uniformly successful), the country-by-country analysis in this paper shows that the performance of individual indicators varies greatly across countries, so that different sets of variables are relevant for different countries. In this light, the one-size-fits-all, panel data approach used in estimating most early warning systems might be one of the causes for their only moderate success. The performance of early warning systems, regardless of the econometric specification chosen, might be improved markedly by taking more care in verifying that the countries used in the estimation possess similar characteristics, or failing that, by estimating the models on a country-by-country basis.

The model presented here is only the simplest variant of what can be done in a Markov-switching EWS. Most obviously, those with a better knowledge of each country can estimate these models using a more informed selection of indicators. Given the role that politics and political stability have played in triggering or exacerbating several crises, most notably Indonesia in 1997, the use of socio-political variables could be explored. In light of increased financial globalization, other external factors in addition to world interest rates might be considered, such as global equity market volatility or the spread on high-yield bonds. Regarding the specification itself, one can extend the model in several directions. First, because the focus was on crisis anticipation, the early warning indicators in the current model only affected the probability of moving from a tranquil to a crisis state. But one could let these same indicators (or a different set of indicators) also affect the probability of getting *out* of a crisis. Second, the current model has only two states: a tranquil state, and a speculative pressure state whose main characteristic is high exchange rate volatility. One could extend the model to allow for three (or more) states, where the three states might correspond to tranquil periods, periods of depreciation pressure and periods of *appreciation* pressure. Finally, the issue of modeling contagion across countries within the context of a Markov-switching EWS awaits further investigation.

References

1. Abiad, A. (2002). "Early warning systems for currency crises: A Markov-switching approach with application to southeast Asia". *Ph.D. Dissertation*, University of Pennsylvania.
2. Aziz, J., Caramazza, F. and R. Salgado (2000). "Currency crises: in search of common elements". International Monetary Fund Policy Working Paper: WP/00/67, March.
3. Berg, A. and C. Pattillo (1999). "Predicting currency crises: the indicators approach and an alternative". *Journal of International Money and Finance*, 18(4): 561–86.
4. Berg, A., Borensztein E. and C. Pattillo (2004). "Assessing early warning systems: how have they worked in practice?". International Monetary Fund Working Paper: WP/04/52, March.
5. Caramazza, F., Ricci, L. and R. Salgado (2000). "Trade and financial contagion in currency crises". International Monetary Fund Policy Working Paper: WP/00/55, March.
6. Cecchetti, S.G., Lam, P. and N.C. Mark. (1990). "Mean reversion in equilibrium asset prices". *American Economic Review*, 80(3): 398–418.
7. Cerra, V. and S. C. Saxena (2002). "Contagion, monsoons, and domestic turmoil in Indonesia's currency crisis", *Review of International Economics*, 10(1): 36–44.
8. Davies, R.B. (1977). "Hypothesis testing when a nuisance parameter is present only under the alternative". *Biometrika*, 64(2): 247–254.
9. Davies, R.B. (1987). "Hypothesis testing when a nuisance parameter is present only under the alternative". *Biometrika*, 74(1): 33–43.
10. Diebold, F.X., Weinbach, G.C. and J.H. Lee (1994). "Regime switching with time-varying transition probabilities". *Nonstationary Time Series Analysis and Cointegration, ed. C.P. Hargreaves*, Oxford University Press.

11. Edison, H. J. (2003). "Do indicators of financial crises work? an evaluation of an early warning system". *International Journal of Finance & Economics*, (8)1: 11–53.
12. Eichengreen, B., Rose, A. and C. Wyplosz (1996). "Contagious currency crises: first tests". *Scandinavian Journal of Economics*, 98(4): 463–84
13. Engel, C. and J. D. Hamilton (1990). "Long swings in the dollar: are they in the data, and do the markets know it?". *American Economic Review*, 80(4): 689–713.
14. Filardo, A. J. (1994). "Business cycle phases and their transitional dynamics". *Journal of Business and Economic Statistics*, 3(12): 299–308.
15. Filardo, A. J. and S. F. Gordon (1993). "Business cycle durations". Federal Reserve Bank of Kansas Research Working Paper No. 28, 93–111, October.
16. Frankel, J. and A. Rose (1996). "Currency crashes in emerging markets: an empirical treatment". *Journal of International Economics*, 41(3): 351–366.
17. Fratzscher, M. (1999). "What causes currency crises: sunspots, contagion or fundamentals?". European University Institute Dept. of Economics Working Paper No. 99/39 (December).
18. Garcia, R. (1998). "Asymptotic null distribution of the likelihood ratio test in Markov switching models". *International Economic Review*, 39(3): 763–788.
19. Goldfeld, S.M. and R.E. Quandt. (1973). "A Markov model for switching regressions". *Journal of Econometrics*, 1: 3–16.
20. Hamilton, J. D. (1988). "Rational expectations econometric analysis of changes in regime: an investigation of the term structure of interest rates". *Journal of Economic Dynamics and Control*, 12: 385–423.
21. Hamilton, J. D. (1989). "A new approach to the economic analysis of nonstationary time series and the business cycle". *Econometrica*, 57: 357–84.
22. Hamilton, J. D. (1990). "Analysis of time series subject to changes in regime". *Journal of Econometrics*, 45: 39–70.
23. Hamilton, J. D. (1994). *Time Series Analysis*. Princeton, NJ: Princeton University Press.
24. Hamilton, J. D. (1990). "Specification testing in Markov-switching time series models". *Journal of Econometrics*, 70: 127–57.
25. Hamilton, J. D. and O. Jorda (2002). "A Model of the federal funds rate target". *Journal of Political Economy*, 110(5): 1135–1167.
26. Hansen, B. E. (1992). "The likelihood ratio test under nonstandard conditions: testing the Markov switching model of GNP". *Journal of Applied Econometrics*, 7: 195–198.
27. Hansen, B. E. (1996). "Inference when a nuisance parameter is not identified under the null hypothesis". *Econometrica*, 64: 413–30.
28. Jeanne, O. and P. Masson (2000). "Currency crises, sunspots and Markov-switching regimes". *Journal of International Economics*, 50(2): 327–350.
29. Kamin, S. B., Schindler, J. W. and S. L. Samuel (2001). "The contributions of domestic and external factors to emerging market devaluation crises: an early warning systems approach". Board of Governors of the Federal Reserve System, International Finance Discussion Paper No. 711.

30. Kaminsky, G. and C. Reinhart (1999). "The twin crises: causes of banking and balance-of-payments crises". *American Economic Review*, 89(3): 473–500.
31. Kaminsky, G., Lizondo S. and C. Reinhart (1998). "Leading indicators of currency crises". IMF Staff Papers, 45(1): 1–48.
32. Kim, C.J. and C. R. Nelson (1999). *State-space Models with Regime Switching: Classical and Gibbs-Sampling Approaches with Applications*, Cambridge, MA: MIT Press.
33. Lee, J. H. (1991). "Non-stationary Markov-switching models of exchange rates: the pound-dollar exchange rate". *Ph.D. Dissertation*, University of Pennsylvania.
34. Mariano, R. S. and F. X. Gong (1998). "Testing under non-standard conditions in the frequency domain: with applications to Markov regime switching models of exchange rates and the federal funds rate". *Federal Reserve Bank of New York Staff Papers.*
35. Peria M. and M. Soledad (2002). "A regime switching approach to studying speculative attacks: a focus on EMS crises". *Empirical Economics*, 27(2): 299–334.
36. Quandt, R.E. (1958). "The estimation of parameters of linear regression system obeying two separate regimes". *Journal of the American Statistical Association*, 55: 873–80.
37. Sachs, J. D., Tornell, A. and A. Velasco (1996). "Financial crises in emerging markets: the lessons from 1995". *Brookings Papers on Economic Activity*, 1996(1): 147–215.
38. Yin, G.G. and Q. Zhang (1998). *Continuous-time Markov chains and applications. a singular perturbation approach.* Berlin, Springer.

Early Titles in the
INTERNATIONAL SERIES IN OPERATIONS RESEARCH & MANAGEMENT SCIENCE
Frederick S. Hillier, Series Editor, *Stanford University*

Saigal/ *A MODERN APPROACH TO LINEAR PROGRAMMING*
Nagurney/ *PROJECTED DYNAMICAL SYSTEMS & VARIATIONAL INEQUALITIES WITH APPLICATIONS*
Padberg & Rijal/ *LOCATION, SCHEDULING, DESIGN AND INTEGER PROGRAMMING*
Vanderbei/ *LINEAR PROGRAMMING*
Jaiswal/ *MILITARY OPERATIONS RESEARCH*
Gal & Greenberg/ *ADVANCES IN SENSITIVITY ANALYSIS & PARAMETRIC PROGRAMMING*
Prabhu/ *FOUNDATIONS OF QUEUEING THEORY*
Fang, Rajasekera & Tsao/ *ENTROPY OPTIMIZATION & MATHEMATICAL PROGRAMMING*
Yu/ *OR IN THE AIRLINE INDUSTRY*
Ho & Tang/ *PRODUCT VARIETY MANAGEMENT*
El-Taha & Stidham/ *SAMPLE-PATH ANALYSIS OF QUEUEING SYSTEMS*
Miettinen/ *NONLINEAR MULTIOBJECTIVE OPTIMIZATION*
Chao & Huntington/ *DESIGNING COMPETITIVE ELECTRICITY MARKETS*
Weglarz/ *PROJECT SCHEDULING: RECENT TRENDS & RESULTS*
Sahin & Polatoglu/ *QUALITY, WARRANTY AND PREVENTIVE MAINTENANCE*
Tavares/ *ADVANCES MODELS FOR PROJECT MANAGEMENT*
Tayur, Ganeshan & Magazine/ *QUANTITATIVE MODELS FOR SUPPLY CHAIN MANAGEMENT*
Weyant, J./ *ENERGY AND ENVIRONMENTAL POLICY MODELING*
Shanthikumar, J.G. & Sumita, U./ *APPLIED PROBABILITY AND STOCHASTIC PROCESSES*
Liu, B. & Esogbue, A.O./ *DECISION CRITERIA AND OPTIMAL INVENTORY PROCESSES*
Gal, T., Stewart, T.J., Hanne, T./ *MULTICRITERIA DECISION MAKING: Advances in MCDM Models, Algorithms, Theory, and Applications*
Fox, B.L./ *STRATEGIES FOR QUASI-MONTE CARLO*
Hall, R.W./ *HANDBOOK OF TRANSPORTATION SCIENCE*
Grassman, W.K./ *COMPUTATIONAL PROBABILITY*
Pomerol, J-C. & Barba-Romero, S./ *MULTICRITERION DECISION IN MANAGEMENT*
Axsäter, S./ *INVENTORY CONTROL*
Wolkowicz, H., Saigal, R., & Vandenberghe, L./ *HANDBOOK OF SEMI-DEFINITE PROGRAMMING: Theory, Algorithms, and Applications*
Hobbs, B.F. & Meier, P./ *ENERGY DECISIONS AND THE ENVIRONMENT: A Guide to the Use of Multicriteria Methods*
Dar-El, E./ *HUMAN LEARNING: From Learning Curves to Learning Organizations*
Armstrong, J.S./ *PRINCIPLES OF FORECASTING: A Handbook for Researchers and Practitioners*
Balsamo, S., Personé, V., & Onvural, R./ *ANALYSIS OF QUEUEING NETWORKS WITH BLOCKING*
Bouyssou, D. et al./ *EVALUATION AND DECISION MODELS: A Critical Perspective*
Hanne, T./ *INTELLIGENT STRATEGIES FOR META MULTIPLE CRITERIA DECISION MAKING*
Saaty, T. & Vargas, L./ *MODELS, METHODS, CONCEPTS and APPLICATIONS OF THE ANALYTIC HIERARCHY PROCESS*
Chatterjee, K. & Samuelson, W./ *GAME THEORY AND BUSINESS APPLICATIONS*
Hobbs, B. et al./ *THE NEXT GENERATION OF ELECTRIC POWER UNIT COMMITMENT MODELS*
Vanderbei, R.J./ *LINEAR PROGRAMMING: Foundations and Extensions, 2nd Ed.*
Kimms, A./ *MATHEMATICAL PROGRAMMING AND FINANCIAL OBJECTIVES FOR SCHEDULING PROJECTS*
Baptiste, P., Le Pape, C. & Nuijten, W./ *CONSTRAINT-BASED SCHEDULING*
Feinberg, E. & Shwartz, A./ *HANDBOOK OF MARKOV DECISION PROCESSES: Methods and Applications*
Ramík, J. & Vlach, M./ *GENERALIZED CONCAVITY IN FUZZY OPTIMIZATION AND DECISION ANALYSIS*

Early Titles in the
INTERNATIONAL SERIES IN
OPERATIONS RESEARCH & MANAGEMENT SCIENCE

(Continued)

Song, J. & Yao, D./ *SUPPLY CHAIN STRUCTURES: Coordination, Information and Optimization*
Kozan, E. & Ohuchi, A./ *OPERATIONS RESEARCH/ MANAGEMENT SCIENCE AT WORK*
Bouyssou et al./ *AIDING DECISIONS WITH MULTIPLE CRITERIA: Essays in Honor of Bernard Roy*
Cox, Louis Anthony, Jr./ *RISK ANALYSIS: Foundations, Models and Methods*
Dror, M., L'Ecuyer, P. & Szidarovszky, F./ *MODELING UNCERTAINTY: An Examination of Stochastic Theory, Methods, and Applications*
Dokuchaev, N./ *DYNAMIC PORTFOLIO STRATEGIES: Quantitative Methods and Empirical Rules for Incomplete Information*
Sarker, R., Mohammadian, M. & Yao, X./ *EVOLUTIONARY OPTIMIZATION*
Demeulemeester, R. & Herroelen, W./ *PROJECT SCHEDULING: A Research Handbook*
Gazis, D.C./ *TRAFFIC THEORY*
Zhu/ *QUANTITATIVE MODELS FOR PERFORMANCE EVALUATION AND BENCHMARKING*
Ehrgott & Gandibleux/ *MULTIPLE CRITERIA OPTIMIZATION: State of the Art Annotated Bibliographical Surveys*
Bienstock/ *Potential Function Methods for Approx. Solving Linear Programming Problems*
Matsatsinis & Siskos/ *INTELLIGENT SUPPORT SYSTEMS FOR MARKETING DECISIONS*
Alpern & Gal/ *THE THEORY OF SEARCH GAMES AND RENDEZVOUS*
Hall/ *HANDBOOK OF TRANSPORTATION SCIENCE – 2nd Ed.*
Glover & Kochenberger/ *HANDBOOK OF METAHEURISTICS*
Graves & Ringuest/ *MODELS AND METHODS FOR PROJECT SELECTION: Concepts from Management Science, Finance and Information Technology*
Hassin & Haviv/ *TO QUEUE OR NOT TO QUEUE: Equilibrium Behavior in Queueing Systems*
Gershwin et al./ *ANALYSIS & MODELING OF MANUFACTURING SYSTEMS*
Maros/ *COMPUTATIONAL TECHNIQUES OF THE SIMPLEX METHOD*
Harrison, Lee & Neale/ *THE PRACTICE OF SUPPLY CHAIN MANAGEMENT: Where Theory and Application Converge*
Shanthikumar, Yao & Zijm/ *STOCHASTIC MODELING AND OPTIMIZATION OF MANUFACTURING SYSTEMS AND SUPPLY CHAINS*
Nabrzyski, Schopf & Węglarz/ *GRID RESOURCE MANAGEMENT: State of the Art and Future Trends*
Thissen & Herder/ *CRITICAL INFRASTRUCTURES: State of the Art in Research and Application*
Carlsson, Fedrizzi, & Fullér/ *FUZZY LOGIC IN MANAGEMENT*
Soyer, Mazzuchi & Singpurwalla/ *MATHEMATICAL RELIABILITY: An Expository Perspective*
Chakravarty & Eliashberg/ *MANAGING BUSINESS INTERFACES: Marketing, Engineering, and Manufacturing Perspectives*
Talluri & van Ryzin/ *THE THEORY AND PRACTICE OF REVENUE MANAGEMENT*
Kavadias & Loch/ *PROJECT SELECTION UNDER UNCERTAINTY: Dynamically Allocating Resources to Maximize Value*
Brandeau, Sainfort & Pierskalla/ *OPERATIONS RESEARCH AND HEALTH CARE: A Handbook of Methods and Applications*
Cooper, Seiford & Zhu/ *HANDBOOK OF DATA ENVELOPMENT ANALYSIS: Models and Methods*
Luenberger/ *LINEAR AND NONLINEAR PROGRAMMING, 2nd Ed.*
Sherbrooke/ *OPTIMAL INVENTORY MODELING OF SYSTEMS: Multi-Echelon Techniques, Second Edition*
Chu, Leung, Hui & Cheung/ *4th PARTY CYBER LOGISTICS FOR AIR CARGO*
Simchi-Levi, Wu & Shen/ *HANDBOOK OF QUANTITATIVE SUPPLY CHAIN ANALYSIS: Modeling in the E-Business Era*

***** A list of the more recent publications in the series is at the front of the book *****

Kolibri 360 GmbH
Ferdinand-Jühlke-Straße 7
99095 Erfurt, Deutschland
produktsicherheit@kolibri360.de